On
Many Paths

Jim Dashiell MD
aka
Funnybone

Naneek

Godspeed along your
chosen path!

Jim aka Funnybone

5-19-17

Copyright 2016
Jim Dashiell

ISBN-13: 978-1540893444

ISBN-10: 1540893448

Indianapolis, Indiana

Introduction

Each chapter in this book is the unique Appalachian Trail story of a different author. A single chapter is not sufficient to relate all the events the contributors experienced on the AT. So, each hiker-author has selected events that were personally meaningful. As you will see, the trail experience means many different things to this diverse assortment of hikers. We all suffered, rejoiced, were disappointed and rewarded almost daily during the many months we hiked through the "green tunnel". This book will encourage you to become more involved with hiking and with working to achieve your personal goals.

I want to thank the Appalachian Trail Conservancy for its dedication to improving and protecting the AT, so that we were able to have the unique experiences described in the pages of this book. And, I thank the ATC for providing the same opportunity to every reader of this book, and all others, who venture out and on to the Appalachian Trail.

Table of Contents

Chapter 1 A Life Changing Experience
Jean Deeds aka Indiana Jean

On a Sunday morning in March some twenty-three years ago, I was relaxing on my cozy porch with a cup of coffee in hand, reading the morning newspaper, when an article caught my attention: A young woman from an Indianapolis suburb had hiked the entire Appalachian Trail, from beginning to end. As she recounted her experience, it didn't sound like fun to me. In fact, it sounded downright miserable! Bugs. Rain. Heat. Steep mountains. Sleeping on the ground. Carrying a heavy backpack, day after day after day. Hiking for more than two thousand miles on a trail through the woods, from Georgia to Maine. Why would anyone want to do that?

I knew very little about the Appalachian Trail at the time, and the term 'thru-hiker' was new to me. I wasn't a hiker, a backpacker, a camper. I'd never done any of those things as a child nor when I was raising my two sons. I liked my activities – a little golf, a little tennis --to end by sundown so I could go home and sleep in a comfortable bed. At age 50, my life was satisfying and enjoyable. I was an empty-nester with one son in college and another beginning his post-college career. My time was filled with a rewarding career as public relations director at the world's largest children's museum, and with many friends and interesting activities. I certainly had no desire to spend six months living outside 24 hours a day alongside my most feared nemesis: mosquitoes!

But as I read the article, I began to wonder what it would be like to take on a challenge of that magnitude -- something so far outside my comfort zone. The challenge was intimidating, but somehow it intrigued me. Could I possibly do something that intensely physical? Was there any reason to even try?

That afternoon, I went to the library and checked out several books about the AT. While reading them during the next several days, I tried to visualize traveling on foot from Georgia to Maine, living in the woods, not knowing where I would sleep each night and whether I would get eaten alive by mosquitoes. But despite many misgivings, before the week was out I had made an incredible decision: I would attempt to thru-hike the Appalachian Trail. Within days, I started planning, buying equipment and training myself for the biggest physical challenge of my life.

A year after that fateful day, I took my first step at the southern terminus of the Appalachian Trail and started hiking north. I went to the trail alone, with the feeling that this was meant to be a solitary odyssey. I was nervous, rather scared, and wondering why in the world I was doing this. But it was too late to change my mind. I had bought my gear, trained my body, quit my job, rented out my house, and made all the necessary arrangements to be gone for six months. Virtually everyone who knew me knew that I was on my way from Georgia to Maine.

The first days on the trail were just as rough as I had expected, but I found that having realistic expectations is part of good planning. I knew it would be the hardest thing I'd ever done, so when it was, that was okay. Tomorrow would be another day and perhaps a better one. If not tomorrow, then surely the next day. Or the next. Fifty years of life experience helped me set realistic goals and withstand difficulties.

But it never got easy. I was one of the older hikers on the trail attempting a thru-hike. Many of my peers were young men with lots of hiking experience who could hike much faster than I could and carry more weight easily. I coveted their strong legs and broad shoulders. But as I made my way slowly up the trail and became friends with my fellow hikers, we were all reaching the conclusion that the challenge was mental much more than physical. It didn't matter much whether you had big muscles and youthful energy. The mental grit and determination were what would get you to Mt. Katahdin.

On the trail, I was afraid of the heights; I hated the bugs; I endured the rain, the cold, and the heat; and my knees ached every step of the way from the weight of the backpack.

Ahhh, but the joys far outweighed the hardships. I reveled in spending every hour of every day immersed in the beauty of nature amid trees, flowers, streams, mountains and stunning views. I loved the freedom of a life that depended on no one other than me, with no schedule, no deadlines, and no pressure to make things happen for anyone else. I felt like I had stepped into an alternate universe. I admired and liked the other hikers I met along the way, with our shared experience creating bonds that were strong and lasting. And the satisfaction that came with being an accepted part of such an elite group of fit, adventurous, goal-driven people was compelling.

Yes, I did make it to Maine, and it was, as expected, the most difficult challenge of any kind I'd ever undertaken. And the journey changed my life. In every way.

When I returned from the trail, many people had read about my journey in the Indianapolis Star, and I became a sought-after speaker. In the ensuing years, I made more than 400 speeches about my journey, I wrote a book, and I took a number of small groups of women to the AT to hike for a week at a time. I also became a part-time tour director with a travel club and was paid to accompany groups on adventure trips around the world. Outdoor adventure came to define me, and I was known as "the hiker lady."

I found that hiking the Appalachian Trail had prepared me for all of these experiences in a way I could never have imagined. Hiking the AT had become a metaphor for my journey through life, with each day challenging me in new ways and lessons about myself emerging along the pathway. When telling my story, in my book or in speeches, I framed it in just that way because my quest had been a daily lesson in living in the moment, not knowing what lay ahead, but dealing with each difficulty or appreciating each joy that presented itself in that very moment. I accepted my fears, but remained confident in my ability to stay the course.

I began to realize that the metaphor in my story was striking a chord with my readers and listeners when I received a letter from a woman who had read my book. In it I had shared the lessons I learned, including the recognition that the only way to make it to the top of the next mountain is to take it one step at a time. And this woman wrote that my journey on the trail mirrored her journey through breast cancer. The parallels were so obvious to her that she felt like I was telling her story. And then an audience member came up to me after one of my speeches and told me that my odyssey on the AT was strikingly similar to her journey through grief after her husband died. Another aha moment. By then I understood that we each learn our lessons in life in our own way, but they are essentially the same lessons. The situations during which we come to understand ourselves and our lives are sometimes chosen by us, but often they are foisted upon us in ways we couldn't imagine and didn't seek. But we survive. Because we have to, until the final step on the final mountain.

Some twenty-plus years later, I often find it difficult to believe that I actually hiked the entire Appalachian Trail. What a gutsy thing to do! I've never done anything else that stretched me so far past what I thought were my limits. I am proud of myself for doing it, and I have been reaping the rewards every day since. I feel blessed every single day for the adventures that have come my way, for the friends I have met, and for the many people for whom

I have been a positive role model. My gratitude is coupled with humility, because I could never have done it alone; support systems were everywhere, in the form of family, friends, fellow thru-hikers and surely at least one guardian angel.

Perhaps most of all, I owe my heartfelt gratitude to one man whose vision of a wilderness footpath winding its way through the rugged mountain ranges of the eastern United States expanded and enriched my world beyond anything I could have imagined. Thank you, Benton MacKaye. *You* are my hero.

~ ~ ~

Jean Deeds thru-hiked the Appalachian Trail in 1994 when she was 51 years old. Public interest led her to write a book about her journey, and she has made more than 400 speeches throughout the country about the life lessons she learned on the trail. Since then, Jean has led groups on hiking expeditions throughout the United States and around the world.

Chapter 2 Appalachian Rabbit Hole
Bonnie Karet aka BonBon

My story about my hike does not start with my first step on the trail. People hike the trail for many reasons, and sometimes the reasons are not even known to the hiker. That was the case with me. I thought I was hiking for adventure and to combat an ordinary mid-life crisis. That was somewhat accurate, but my journey was so much more complicated than that. The trail revealed my true path early on. It was a hike up and over mountains, through forests, across sparkling streams, and through small Appalachian towns. It was also a journey in and out of an obscured rabbit hole, created by my own previous adventures.

My rabbit hole begins in my childhood. I think that is true for all of us. Because of our experiences, we learn to bend as we grow in strength and flexibility, or, we break as we allow life's lessons to make us brittle and fragile. Ernest

Hemingway said, "The world breaks everyone, and afterward, many are strong at the broken places." There were parts of me that were broken when I decided to hike the AT in 2015, and the hike would reveal to me the exact nature of those breaks and what I needed to do to fix them. Unacknowledged damage to our souls is cumulative and results in baggage that we carry around with us, ever impacting our relationships and our happiness. All my experiences ultimately led me to the Appalachian Trail, that long and judgment free path, where I could finally bleed and heal.

I grew up overseas, in mostly countries that Americans are now discouraged to travel to; India, Greece, Pakistan, Philippines, Saudi Arabia, and Liberia. My Dad worked at the American Embassy and we often lived on Embassy compounds or nearby residential compounds. All of us were expected to adhere to the rules; stay inside the walls, don't mix with the indigenous population, and don't be a trouble maker. I broke all those rules beginning at a very early age. When I was six years old, I joined the little street urchins at the market near my house in Rawalpindi. I hung out with them while they sold combs, sunglasses, and gum. I sold my mother's cigarettes, which she bought from the PX. I stole them from her by the carton. The gang of children I joined were dark skinned with beautiful silky black hair, wearing dirty and tattered clothing. I was pale and tow headed wearing nice clothes from the Sears catalog. I yelled out "Cigarettes! American cigarettes!" I remember it still, the smells and the sounds, the feeling of connection and belonging. Our cook would come searching for me and drag me home by my ear, scolding me in Urdu.

I remember often looking out from the gates that enclosed the compound, gazing past the snake charmers and milk carts at the streets and city beyond, imagining how delicious life would be if I could escape. As I looked out, impoverished children gripped the iron bars and looked in. They were mesmerized by the pruned shrubs, the beautiful homes, and the clean scrubbed white children laughing and playing. How wonderful life must be inside the gates they were thinking. The kids inside felt like we were living inside a cage. My cage was gilded, for sure, and their cage was one of severe poverty and futures that did not include education or prosperity, or even peace in their country.

In the years that followed, in all the countries I lived in, I became a very skilled wall jumper. I could evade marine guards, scale walls embedded with glass shards effortlessly. I was fearless and curious and never allowed a little punishment (if I was caught) to hold me back the next time I wanted to roam. I savored the sweet and perfect freedom I felt when I hit the ground and disappeared into the busy streets. I fell in love with each place I lived and bonded with local and embassy kids. And after a couple of years, we would move to a new country. With each move, each heartbreak at leaving friends and house cats behind, my baggage grew a little. I didn't know it though. I loved my nomad life.

Returning to the States to start high school was not an adjustment I made well. I had a bad case of wanderlust and I just couldn't adapt to life in what I perceived to be a bland land, filled with colorless, odorless people. There were no walls to jump and no marine guards to sneak past, but I ran away from home frequently, hitchhiking all over the country.

I remember vividly the feeling of standing alone on a deserted highway, sky black and quiet. As the headlights of trucks approached, I would stick my small thumb out into the big night, really testing the fates. I loved that feeling, not knowing what was next or how I would react. My perfect freedom.

All the bad things that you imagine could happen to a 15-year-old girl alone on the highways happened to me. I grew my baggage, creating the cracks that would lead to the breaks that would lead me to the Appalachian Trail some 35 years later. I stuffed the emotions, the anger and the hurt inside my baggage and locked that shit down. I just marched blindly forward, denying myself the space to grieve the innocence that I had so ruthlessly given away.

When I was in my early twenties, I worked in the surveying department of an engineering firm in Northern Virginia. I had read a riveting account of a thru hike of the Appalachian Trail. Ever in my imagined cage, I yearned for the experience. I was young though, with no savings. I needed help to pull it off. "See the Appalachian Trail through the eyes of a lone unarmed woman and her dog," proclaimed my small ad in a newspaper. I promised potential donors pictures and letters from the trail. I was a trail-blazing adventure-beggar, something that has been polished by sites like GoFundMe in today's world. I did not get a single hit on that ad. I waited with growing disappointment, but my mailbox yielded none of the flood of anxious folks seeking a vicarious adventure that I had imagined. So, I worked my job, went to school, and did normal twenty something things. The AT dream receded and then just sat dormant in my brain for the next 3 decades.

In my mid to late 20's, I loved a man and lost him to a violent white water drowning. I moved on with our baby son and found love again a few years later, stuffing the agony of that death into my baggage and putting a bigger lock on it. My new husband and I added two daughters to our family. Life just plugged away. I was homeroom mom, team mom, avid runner mom, and business owner mom.

The little glowing nugget of the AT dream, squirreled
away in some dark corner of my psyche, waited patiently
for the sunshine. The nugget started to fester when I hit
my late 40s. The locks on my baggage were starting to
rust a little. The AT dream burst out of my brain and into
my actual life in my 50th year. The resurrection of this
idea came at a time when I was struggling in my life. It is
no surprise that folks have a "crisis" at middle age. It's
the Harpers Ferry of our lives -- the emotional halfway
point. When I turned 50, for the first time in my life I
experienced deep reflection. I was married to a wonderful
man and I owned a successful business. We had three kids
we loved dearly. On the surface, everything was terrific.
But I was acutely aware of the path that had led me to
where I was, the choices I had made, the secrets I had
kept. I experienced that gnawing self-doubt that grows
inside when you think if everyone truly knew you, what
you had done, they would not love you anymore. I saw
my past very clearly but with a certain disconnect. I
couldn't see my future.

Being a lifelong athlete, I had inexplicably and quite
suddenly quit exercising at about age 45. I started
drinking. And eating cheeseburgers more than once a
week. I gained 50 pounds in just 5 years. I allowed stress
from my business to overwhelm me. Because I was not
exercising, I had no endorphins to help ward off the blues.
I felt fat and unattractive. That spirited runaway that had
guided -- and misguided -- my life was leaving me. I
knew it and I mourned her. I was turning into a stressed
out and negative person as my spontaneity and joy
drained away.

My conscious decision to hike the entire Appalachian
Trail was to combat the middle-aged apathy that was
taking over my life. I wanted to reclaim my fit self and
my inner rebel. I wasn't ready to give in and be a couch
dweller. The unconscious part of the decision, which I

believe is what truly compelled me to do this, was a profound survival instinct that had been asleep for too long.

During the year leading up to my hike, my husband endured my endless exaltations of things I learned about the Appalachian Trail. Once I fix on an idea, I am the proverbial dog with a bone. The information I gathered maniacally for the year leading up to the hike and shared daily with my husband was probably more than a little annoying, but my man is solid. He listened, he nodded, he said, "You don't say." And, "Is that a fact?" I whittled away hours walking a trail near my house, doing important things like turning over potential trail names in my head. I followed a few journals on *trailjournals.com*: Left Turn, Red Robin, and Affirm were my favorites. I bought gear they liked. I pictured myself in their situation of the day. I was wholly and obsessively immersed in the trail culture before my boots ever touched the earth of Georgia. I walked the flat and unchallenging trails of Florida, adding more and more weight to my pack each week. The months to wait turned to weeks turned to days until the hike was to begin. My excitement grew.

Because I was overweight when I started, and had never backpacked before, people secretly believed I wouldn't make it. I mean, the odds are not good for young fit people, so for me they were even worse. I knew people were thinking it, maybe even snickering about it, but nobody said it out loud to me. I never doubted I would make it. In my mind, I was still the athlete I had been all my life except the last 5 years.

I had met another hiker through trail journals named Katwalk. She had this crazy plan with her hiking partner to rent a cabin for a week and shuttle each day back and

forth from the beginning and ending points. The goal was
to ease into the hike. Her partner dropped out days before
the hike was to start. So Katwalk invited me to join her.
At first, I didn't want to. It seemed less than authentic to
me. I ran it by my husband and told him I wasn't going to
do it. He said "Are you crazy? Of COURSE you should
do it." It sounded very smart to him. I accepted the offer
and good thing I did. I look back on that decision as the
best decision I made on my hike.

My husband and daughter dropped me off at the Tampa
Airport on March 22. I had my lock blade taken from me
at the security checkpoint, but they left me my bear spray.
I had at least 50 pounds of stuff in my backpack and a
duffel. Most of my gear still had price tags on it. The only
thing I had practiced setting up was my tent.

When we landed in Atlanta, I carried all that stuff up the
jet way. The incline made me gasp for air and I stopped to
rest at the top. It was in that moment that reality
introduced it's snarky self to me. I stood there, breathing
hard, and acknowledged to myself that the jet way was
not a mountain and I might be in a little over my head. I
threw my head back and laughed. I like not knowing
what's coming next. It felt like I had come home in my
head. Katwalk and I picked up a few bottles of Two Buck
Chuck and some food stuff at Trader Joes for our cabin
stay. We bonded over our mutual love of Jane Austin
books. I didn't know it then, but my friendship with this
free spirited and generous woman would end up being
one of the most precious gifts of the AT.

On March 23, 2015, I hauled my fat ass up Springer
Mountain to begin my hike. It was my wedding
anniversary. I chose that day to start because it is a good
day for me regarding long term commitments.

My heart was beating out of my chest. As I hiked up that first incline to the Springer Mountain summit, I kept thinking, "I'm really here -- I'm really doing it." I struggled up that mountain. I could have skipped up it 6 months later. Mica sparked and shined up from the trail, each tiny glint a promise of the glory yet to come. When I stood at the summit with Katwalk, we reveled for a moment in the beauty, breathing deeply the cool clean air. I gazed out at the view and committed myself to the hike, to love and to honor till Katahdin do us part.

Katwalk and I used that first week as a shake-down hike. We tested our gear and our legs. I learned how to use my water filter, how to use the bathroom in the woods. I practiced my Leave No Trace techniques. I attempted to hang my food bag in the trees so bears could not get it. This is maybe the only task in my entire life that I cannot master. From the first day to the last, I was a menace hanging that food bag. I nearly concussed countless people with the rock bag and never got the hang of it. Katwalk and I avoided the overcrowded campsites during what was the busiest start-week of the hiking season, and went back to the cabin each night. We drank our cheap wine and grilled steaks and talked about the amazing things we had seen that day. We met some famous trail folk -- Miss Janet, Pirate, and Baltimore Jack. Katwalk dubbed them Trail-ebrities. We were generally giddy in this early phase of the hike, all our AT dreams seemingly becoming a beautiful reality. We only carried about 20 pounds in our packs that first week to give our bodies the chance to ease into the hike.

On the day we were to leave our cozy cabin, we finally packed all of our gear in our packs. Neither of us could hoist the packs onto our backs. We spent an hour eliminating things we probably wouldn't need and both unloaded about 10 pounds. We went to Mountain Crossings at Neel Gap to weigh our packs. Unfortunately,

it was a floor scale and part of my pack rested on the
ground. The needle said 31 pounds. Ignorance of the true
weight was not bliss. I put what was closer to a 50-pound
pack on my back and headed in to the mountains. From a
distance, it probably looked like a small man was on my
back trying to take me down.

Our first night camping was at a shelter. We set up our
tents, mine an orange Big Agnes and Katwalk's a blaze of
fuchsia that must have been visible from the space station.
We walked down to the shelter and sat at the picnic table
and I attempted to read the tiny little directions for my
cook stove. I figured it out, with the help of a German
hiker. We retired to our tents, so very excited to sleep for
the first time in the woods.

A violent lightning and rainstorm kept me alert and fairly
terrified for most of the night, but I stayed dry. Katwalk's
tent failed, and she headed to town to dry everything out.
I continued North and spent the first night all by myself in
the woods. I cooked my meal, spent a considerable
amount of time hanging my food bag, and experienced
my first fall in my camp shoes. They were a pair of bright
orange crocs I got from the Goodwill, about two sizes too
big. I caught myself just before my face made contact
with the slab of granite I was headed for. Tripping and
falling would be a regular occurrence for me. That night
in my tent, all alone in the big dark woods, was the most
terrifying and exhilarating night of the entire hike. Every
snapping twig, every rustle of leaves, had me in a near
panic. I had to get out of my tent twice during the night to
pee, and each time I peered through the screen into the
darkness looking for bears or Zombies. The Moon chased
the Sun, and the moon-glow, undiminished by artificial
light, cast an eerie and beautiful wash of muted color for
as far as I could see.

Katwalk and I hiked another week or so together, and then, as often happens, our pace separated us. I hiked on alone. I met a couple from Florida, D.No and C.Shell, a few days later. We leapfrogged each other for a few days and ended up at the same time at the Nantahala Outdoor Center. C.Shell and D.No shared a meal with me and then Katwalk came down the mountain too. It was a very nice night.

Later, I walked to the river edge and put my feet in the water. It was icy cold and felt wonderful. I watched the kayakers and thought about Brian. He died in a whitewater kayaking accident nearly 25 years ago when our son was only 9 months old. A charismatic adrenaline-junkie, he would have loved it here.

The next day, I had a nearly 3,000 foot elevation gain over nearly seven miles. It was a long, tough climb. I was inside my head, and my head was going places I had avoided for a quarter of a century. I thought about Brian, about our tumultuous relationship. What he said, what he did, what I said what I did. I cried, I raged, I totally went down the rabbit hole that opened while I watched the kayaks the day before. I let the shame of my own immature behavior in that relationship wash over me, owning it, finally. His death had robbed us of our resolution and I had carried that guilt with me until I laid it down that day on the trail. When I reached Sassafras Gap Shelter, I was totally spent, emotionally and physically.

Later in the shelter, I was drifting off to sleep when I heard a new voice join in the campfire conversation. A soft-spoken guy had come down to the shelter, his trail name was SOBO HOBO. He was nearly done with a southbound thru hike. He told stories and gave advice. His voice was mesmerizing and I had to get up to see who this person was. He was lean, quiet, sage-like. I felt like

17

we were children of a lesser god compared with this guy. He never hiked in the rain. Cool. I can embrace that advice. I wondered when and if I would morph into a creature like him. I was destroyed by my hike that day, and this guy stayed a while then got up to leave, in the black of the night. We asked where he was going. He said, "NOC; it's not far, right?" Right on, brother. Hike on.

A couple of days later, I started the Smokies. My entire time in the Smokies was heralded by cold driving rain. I experienced "misery of the acutest kind", as Elizabeth Bennet would have said. But, I also started hiking with D.No and C.Shell. We met up the first night in the Smokies, cold and wet and were so happy to see each other. Because our paces were the same, our time together lasted over 500 miles, and we would separate for 500 miles then meet up again in Pennsylvania and hike another 500 together. It was a partnership that was so precious to me, and only thru hikers know how intense the relationship of "trail family" is, how rewarding, how hike saving.

My brain continued to purge old wounds and hurts throughout the hike. The experience of being alone in my head allowed me the space to honestly see my own role in all the perceived wrongs that had been dealt me. The awareness I experienced was more painful than the discomfort of walking up and down mountains for 10 hours each day, falling, twisting ankles, walking with blisters. With each revelation, each honest acceptance of my own mistakes, I felt lighter. I dropped bits of my baggage on that trail and it swallowed them up -- always without judgment. I reached back through time and embraced the confused child that I had been. I gave myself a break. As I hiked, my body became lean and strong. The camp chores that had baffled me in the beginning became as natural as breathing (except for

hanging the food bag). I came to a unique place of being
completely raw and broken down emotionally while being
the strongest physically that I had ever been. By the time I
reached Harpers Ferry, my body was 25 pounds lighter,
my pack was down to 35 pounds, and the weight loss of
my emotional baggage was unquantifiable.

In Harpers Ferry, I spent time with my husband. He and I
explored the town, held hands, laughed and enjoyed each
other. He never let me feel guilty about leaving him and
our 15-year old daughter. That support on my hike was so
valuable.

I hiked North out of Harpers Ferry. The terrain had
become relatively easy, compared to what I had left
behind. The trees were now lush and full, and the weather
was hot and muggy. In Pennsylvania, I met up with my
daughter and spent a few days with her and with my
brother and his wife. Zoe was sweet and supportive, and if
she cried when I walked away from her a few days later,
she didn't let me see. I cried and did not want to leave
her.

My brother was my hotel room trail angel. He had
literally a million rewards points with Holiday Inn
because of his business. If there was a Holiday Inn within
10 miles of a road crossing, I would be there. I stayed 13
times in one of these rooms, bathing, sleeping in cool
white sheets, and washing my mephitic backpack.
Sometimes alone, sometimes with my hiking partners, we
reveled in the luxury.

I met back up with D.No and C.Shell in Pennsylvania. We
hiked through the easy states together. We were happy
and laughed a lot, our friendship growing. In mid
Vermont, the trail became increasingly difficult and
beautiful. Is it my imagination or do beauty and pain walk
hand in hand? Descending Mount Killington, I was

almost struck by lightning. It exploded directly in front of me. The rain was coming down sideways and the trail was a torrent of rushing water. The lightning strike was so intense one of my hiking poles flew out of my hand. Over the next few days, burns and capillary flowering appeared on my body. Everywhere my pack touched me was bright red, then blistered and bled. I hiked on, ears ringing for months after that.

Katwalk had gotten off the trail at about 600 miles in because of an injury. She appeared one night as a glowing angel in the form of a text on my phone. I whined to her about how hard the trail was, how heavy my pack was. She offered to loan me her ultra-light gear for the remainder of my hike. With the Whites coming up, I was tempted, but I knew I would stink up her stuff so I declined at first. She insisted and mailed her stuff to a Holiday Inn on my radar. It lightened my load by four pounds and I felt as though I could fly!

D.No, C.Shell, and I loved to "stealth" camp. We found secluded spots most of the time, avoiding shelters and campsites. The nights spent gazing at stars or reading in my tent were so spectacular. The trail got tougher and tougher, and finally we reached the Whites. This mountain range is brutal. It so breathtakingly beautiful that it can distract you from how hard it is, but it is one challenging climb after another. We trudged through, alternately cursing and thanking the gods that brought us there. We hiked in the dark, took nasty falls, and each of us suffered injuries. We wondered why this was fun. D.No's knee became so bad that he hobbled, but he refused to stop. The day we reached the sign that welcomed us to Maine, we brought it in for a group hug. Georgia to Maine. We were feeling very powerful. Hiking in Maine was like turning a page. It was still very difficult. The terrain had not let up, but were so glad to have the Whites in our rear-view mirror.

Hiking in The Whites was like being in an abusive relationship with an extraordinarily good looking person. Maine is more George Clooney; rugged, good looking, and fun. The Whites were like a skin-tight pair of uncomfortable white leather pants -- Maine is your favorite pair of broken in blue jeans. The Whites are Veronica, Maine is Betty. The Whites are a sandwich made of glass shards and a rare delicious jam; even as you spit out blood you exclaim how wonderful it tastes. Maine is a strawberry rhubarb pie. But I digress....

Maine is lovely and earthy and so green. I saw rocks covered with moss, thin here and there, reminding me of well-loved velveteen toys. Foliage is thick, almost jungle-like in some places. When we got to Rangeley, the journey was over for my hiking partners. They had hiked nearly 2,000 miles and D.No's knee finally gave out. I hiked away from them with tears in my eyes.

The day I hiked to West Carry Pond, I passed lots of day hikers. They must have known by my hairy legs and smell that I was a thru hiker. They congratulated me as I passed by them and told me different stories of the terrain ahead. It's hard. It's easy. It's flat. It's mountainous. They were all correct because like everything else in this life, we perceive things through our own filters.

I stopped for a break to read emails. No good news about my business, which I had left in the hands of my extremely capable staff, not one of whom was over the age of 20. We had a good summer but now the dreaded September was upon us. It is always the most painful month of the year. I should have been home two weeks earlier and I still had two weeks to go. I felt I had crossed the line of where people were supportive of this hike to where I was hiking at the expense of others. I could feel the toxic tendrils of stress reaching for me, still there, waiting not so patiently for me to return to my life. As I hiked, with nothing to do but put one foot in front of the other and obsess, I considered just leaving the trail at Caratunk and going home. It felt unfair to let my employees stress anymore and endure all the other stuff that comes with a bad month in retail.

Eventually, the trail led to a postcard perfect lake, bathed in serenity. I stripped off all my clothes and went skinny dipping. The water felt great. I pitched my tent next to the water. As I cooked dinner, I watched loons play in the water. They swam joyfully and noisily back and forth, chasing each other.

I sat on the rocks next to the lake, as the Sun I couldn't quite see set and made the water come to life with color and light. I ate my dinner looking at the incredible beauty around me. And then this heathen decided to pray. I wanted to ask the beautiful earth in front of me -- to which I now feel a profound connection -- for clarity, for strength, for wisdom. Instead, I was hit with a wave of gratitude so strong and so sweet that I long for it still, and seek it. I decided that I would not go home and be a victim of stress, an unhappy person who gets overwhelmed by a life of my own choosing. I fell asleep, listening to the haunting call of the loons.

There is a saying out here: The trail provides. The next day as I hiked, I thought about all the ways in which this had been true for me. I always got what I needed (not necessarily what I wanted) at exactly the moment I needed it most. Sometimes I didn't even know what I needed until I had it. It could be something tangible, like water left under a tree, or something else, less defined; a triggered memory, a sight in the woods that stirred me, or a gentle rebuke if I was being an asshole.

Since the beginning of the hike, I had wild strands of thought and emotion shooting out in all directions and I couldn't make sense of any of it. Ambition (oh, I always wanted to finish), excitement, the emotions and ah-ha moments that sweep through you as you turn everything you have ever experienced over in your head while you walk and walk and walk. Walking over 2,000 miles had worn me down, taken me to a state of being raw and open. Now as I neared the end, those disjointed thoughts, feelings, and revelations were coming together. I thought I could see everything. Where I was, what is important now, where I was going, and how I was going to get there.

A week later, I was at the last shelter I would camp at on the trail. I had made it through the incomparable 100-mile Wilderness. The shelter was populated by thru hikers finishing, southbound hikers starting, and section hikers. I was cooking my dinner and thinking back to the climb out of the NOC over 2,000 miles ago, when Sobo Hobo had drifted in to the shelter. I was so mesmerized by his lean hikerness and wondered if I would ever attain that aura. Tonight, as I whipped up my dinner and ate it in the time it took the brand new hikers to find their stove in their gigantic packs, I thought, yes. I am a thru hiker. I stink, I'm skinny, I do these camp things without thinking. I felt calm and connected, confident, and like I belonged on this trail. The transformation was complete. I had come a long way from the 50 pounds overweight-woman with price tags on her gear standing on Springer and winded from the little climb up.

I hiked to the beginning of Baxter State Park the next day, and caught up to Katwalk, who had flown up to hike with me. We are never really alone, if we allow others in. My hike taught me that. Katwalk and others who care about me proved it. I hiked to the top of that famous mountain, Mount Katahdin, on September 28. I did not cry. I celebrated with many of the people I had met and shared this journey with and felt a deep sense of accomplishment.

And then Katwalk and I got on a plane and headed back to Florida. I watched through my window as the terrain it took me six months to hike slipped by in just three short hours.

I would love to tell you that when I came home, the truths that were revealed to me on that hike "fixed" me. They didn't. They just opened me. I fell down that rabbit hole further and faster when I got home. Immersed in the reality of my day-to-day life, the pledge I made to myself by the lake that brilliant night was forgotten. I experienced the post-hike depression many others have described. I gained weight. I started drinking more than I had before the hike. Stress began its insidious strangulation once again.

When Funnybone asked me to write a chapter for this book, I started reading my trail journal, from start to finish. I was astonished at what I had written. I did that, I said to myself. I was so surprised and moved by my own words, my own transformation, and yes, my own strength. I realized that I was headed down a dark path that was very familiar to me. Back down the rabbit hole. I made a list of things to do -- one at a time -- to relieve the stress in my life. Then I started removing them. The business -- I sold it. The drinking -- I quit. I started walking again. The honesty and the revelations I experienced and journaled about on the trail helped me out of the rabbit hole. This time slower, this time with more intent.

I have finally found my perfect freedom, and it turned out to be freedom from my own self judgment.

I am comfortable in my home, sleeping between sheets and showering daily. I have all the luxuries of a Holiday

Inn Express. And yet, not a night has passed since my return that I have not longed to be looking through the rain fly on my tent. I will never forget the trail. It is a gift that will forever open.

~ ~ ~

First, I am a wife to Mark and mother to Gage, Mackenzie, and Zoe. I have been a business owner since 2004 and recently sold it. I am the co-founder and VP of Big Big World Project, a non profit that puts kids at a small orphanage in Vietnam through college. I have travelled each year to Vietnam to visit these special kids

since 2008 and am so happy that we have 5 college grads and 4 in college now through our program. My hike was a fundraiser for them. See what we do at www.bigbigworld.org.
My family has just moved back to the Daytona Beach area and I am mulling over what is next for me. My life ahead looks like a big blank canvas and I have not decided how to paint it yet.

Chapter 3 Hiker From Down Under Comes Out on Top
Tanya Bajor Blakeman aka Gipcgirl

Aussie Hiker Beats the Odds and Summits Katahdin

I'm Gipcgirl, I'm 66 years old, a hiker and this is my story.

I have white hair, note white not grey, note not grey, white. Its long and I braid into pigtails two of them one on either side of my ears, like when I was six years old. What's wrong with me? This is unacceptable in the real world, its supposed to be short and dyed brown or black

and its supposed to be groomed. Groomed? What's that, I
ask. Groomed! What, like a poodle? Nah, I'll stick to the
pigtails, low maintenance. I need low maintenance
because I am a hiker, a thru hiker. You know, a pack on
my back, hike miles and miles, live in a tent. Oh! You
don't know? Well, never mind. You soon will, if you
keep reading my story.

This madness started a long time ago. Well before I
turned 66, it started one morning in 2008. I had just
awaken and was getting ready to go to work. I suddenly
thought I don't have to do this work thing anymore, I
have enough to get by, I'm tired of working, being polite,
ever so polite to my clients, bowing and scraping, yes sir
no sir three bags full. By the time I had picked up a take-
away coffee and arrived at my place of work 25 minutes
later it was a done deal. My mind was made up, close my
business and put my house up for sale. I was in retirement
mode, what was I going to do, sit back relax, grow
veggies, knit sox? No, I was going to go on a long
distance hike, 2,184 miles to be exact, or 3,519
kilometers. The shock/horror of family and friends, the
questions and the disbelief of the answers: Oh, so you are
just going to get up and go? Uh huh. By yourself? Yup.
Alone? Yes. Where? America. On your own? Yes.

I am going to hike the Appalachian Trail, the AT, all by
myself. Yes, I know it sounds crazy. I'm excited, I'm
obsessed, I start reading books written by other hikers, by
hikers who have succeeded or not. I think about nothing
else. I start gathering the gear I will need. Everyone
around me thinks I'm deranged, but my mind is made up.
I apply for a visa entry to the US of A. The internet
search "USA visa" -- Wow! I can get a 90-day visa; no
effort, fill in the form and bingo I have a visa. Wrong!

I don't want 90 days; I want, need, at least six months.
Hmm. That's not so easy. I have to have an appointment

for an interview. Three days later after lots of phone calls to many different departments of the USA-consulate office in Melbourne (I live in Western Australia; that's the other side of the country), I get an appointment, 8.30 a.m. on the 17th November. I'll need to fly to Melbourne and bring documentation -- usual stuff, passport, proof of money in the bank, proof that I have reasons to return to Australia if I should be granted a visa. Okay, a bit odd but yes I can provide all that proof. After all, I still own a home, it has not sold, yet!! Yes I have children, what? Bring their birth certificates. Now that's getting too weird; they are not children anymore, they are adult men. Oh! I get it; I will return because they live in Australia, really!!

Interview day arrives and I'm at the American Embassy in downtown Melbourne at 8.15 a.m. in plenty of time. Who are all these other people sitting around looking bored. They also have interview appointments for 8.30 a.m. There is at least 30 of us. So I stand in line clutching my folder with all the paper work, as well as my handbag slung over my shoulder. As the line progresses I note that handbags, computers, phones, everything except documents are given up and in return we get a number. Mine is 68. We are assured all possessions will be safe. I feel naked, so do the others as one by one we are searched with a hand held device that sweeps our bodies up and down under our arms between our legs. We are then ushered upstairs six at a time in a lift with two security guards. I have no idea which floor we are on as we arrive at our destination. Here is another waiting area with more hopeful visa candidates all waiting a turn. There are now around 50 or so, mainly young people, waiting. One by one numbers are called out in, what to me seems like random selection, as they are not sequential. So, I have no idea how long this will take.

I have now been waiting here for three hours. I should not have had two coffees before arriving for the interview. Others are also getting fidgety — I really need to go pee. The little group around me was discussing this when a little white-haired lady gets up and goes to talk to one of the security guards. After what seems like a long chat, she goes through the exit door. We wait and wait for her return. Nearly an hour goes by before she returns and tells us that she had to go back downstairs and go to the restroom, then return to the line to be taken through and had to be searched again. No toilets on the interview room floor. I change my mind. I don't need to go anymore.

At long last my number is up on the glass wall almost 5 hours after my arrival. I am bored, tired, irritated, and I still need a pee. I am also very hungry. I hand over my documents. In silence the man looks at them and says to me in a very strong American accent. "Why do you want to spend six months in the USA?" I want to say that after this morning I'm not sure I want to go anymore. But, probably not smart to say that, so, instead, I say that I want to hike the Appalachian Trail and it will take me 6 months or longer. Its 2,184 miles. Suddenly, he's all animated. He is a thru hiker -- bingo! What luck for me. He thru hiked with his Dad when he was 17 years old. We chat about the trail and he wishes me good luck, stamps my passport, and tells me my visa will be in the mail.

The day has arrived. I pick up a registered envelope from the post office. I have received a multiple entry visa good for six months in a given year for the next five years. I am so happy. I have no idea how this kind act from a visa issuing officer will change my life forever.

Preparations start in earnest. My spare room becomes AT headquarters. Three piles of gear: must have, maybe, and just in case. I have no idea how all this is going to fit into my pack, a 65-litre pack that weighs three pounds empty.

I don't take any notice of "light weight" or "ultra lite";
these words are foreign to me so they are ignored. After
all, I have hiked in Australia.

The house is sold. I need to pack stuff, sell stuff, get rid of
lots of stuff. 30 years of hoarding stuff and now my life is
reduced to a 3 x 3 meter storage shed and a backpack. My
flights are booked and eventually the day of departure
finally arrives.

The drive to the international airport is 295 kilometers
from where I was staying with friends. My "tin tent" on
wheels goes into long term storage and I'm finally on the
airplane. A stop over in Melbourne to catch up with my
disbelieving sons -- they still think I'm crazy but are
resigned to the fact that I am going. "We will miss you,
and stay safe!" Its a long and very boring flight watching
movies, eating plastic food, and trying to sleep. The
negative doubts about what I'm about to do go around and
around in my head, maybe I am too old and not so crazy.

Arrival at LAX is chaotic. I'm wandering around, filling
in forms, standing in line, all of us tired and cranky. Two
other planes arrived at the same time. Border security
staff are also cranky, bossy, not polite or friendly, in fact,
very bossy. My turn and all is good. I'm finally in the
USA. Eventually I find baggage claim, but not realizing
that I should have caught a train, I walked down a
concrete tunnel-like path wondering where everyone had
gone. I find my pack, the only piece of baggage left on the
round about.

Its 6:00 a.m. I manage to find the train I need to catch and
think to myself: Get to North Springer Station, sit and
relax, eat lots, have coffee, and wait for the hostel staff to
pick me up. Wrong. North Springer Station is a concrete
jungle -- nothing there, no food, no coffee. But I'm too
tired to go anywhere else. Its freezing cold. When I left

West Australia it was 40 degrees Celsius, which is around 100F, and it's now maybe 30F or lower.

I'm freezing. I dig out my Puffy jacket, my sleeping bag, and huddle down behind a concrete wall and wait. I have faith someone will arrive. They know I'm coming. I have no phone, so can't ring them. Five hours later and frozen solid my ride finally arrives. Some communication mix up. That's Okay, I say. I'm so glad you are here.

The hostel is awesome, clean and tidy. I share a room with three others and meet lots of hikers preparing to start the AT. The next day is filled with food shopping, meeting more hikers, eating pancakes for breakfast, and a pack shake down. My pack weighs 40 pounds. Hikers are full of good advice, mainly saying, "Way too heavy, get rid of half that stuff you, won't need it." Trust me, I do need it. I have hiked a lot and always carry roughly the same amount. Aussie's are not into ultra lite. We don't hike as far each day and we go much slower.

I will regret those words and by the time I eventually get to Hot Springs, I decide to shed the weight and buy an ultra-light ULA pack. I change most of my gear to get my pack weight down to 25 pounds. Wow! What a huge difference. I feel like a bird, weightless flying over the trail. In fact, running in some places. True freedom. Oops, I have jumped ahead by many weeks. I should get back to my first day.

Day 1, 19th March 2010. We get dropped off at Springer Mountain. A short day, and I meet many hikers. We are all in high sprits. Day 2: I'm an early bird and started hiking at 6.30 a.m. It's very cold, but it's okay. I'm outdoors, I'm excited, scared, happy, and just want to be moving. The day is beautiful and the woods are quiet so quiet. The bush in Australia is noisy with parrots screeching, kangaroos thumping, and many other loud sounds. The days roll on, one day much the same as the next, hiking many miles and meeting many other hikers.

Day 6 is disaster day. It starts off very early. It rained the night before, so the ground is wet and very muddy. I get to the top of Tray Mountain, and as I start down the other side I stumble on a rock. Incredible pain in my left ankle as I fall four meters down the mountain side coming to a halt when I hit a tree. There I lay covered in mud and in agony. This is serious. I crawl back up onto the trail. Taking stock of my condition, my head says I've broken something but my heart says I'll be okay. Nursing training from many years ago springs into action. Leaving my boots on I slide sticks down each side of my boot. I dig out of my pack a bandage. Using duct tape I tightly wrap up my ankle and half way up my leg. Instinct tells me I have to get down off the mountain to the road, so crawling and sliding on my backside down the switchbacks I eventually get down to the road. Fortunately, I get help and a ride into Hiawasse.

I have broken my ankle and end up with my leg encased in a heavy black walking boot. No hiking for six weeks. It was recommended that I go home, back to Australia. No way. I decide to wait for the ankle to heal. I'm in Franklin, a nice town to hang out in for the duration. The town folk adopted me and I was never short of a home cooked meal or someone taking me out for a drive. I even managed an overnight camping trip. During the time spent in Franklin I met so many excited hikers off on their amazing adventure. I wondered how many would actually finish, and whether I would see them again.

Exactly five weeks after my accident I was ready to hike once more. Initially I could only handle eight or so miles a day, but each day I was getting stronger. It wasn't long before I was hiking 18 or so miles a day.

Hiking in America is so different than in Australia. The main reason is people. To me, the number of starters on the AT is staggering. On most long distance trails in Australia, unless you are in a guided tour group or start off with a partner, you would be lucky to meet any other hikers, maybe the odd ones at a shelter. I have hiked for days and days, even weeks, and not seen a single hiker. Also, we don't have trail magic. I was amazed at the amount of hikers, ex hikers, family of hikers, organizations, etc., that help out hikers on their journey.

The first time I encountered trail magic was so unexpected; food, drinks, chairs, great company, and lots of encouragement. Many times on my hike I would be so tired, dirty, hungry (I was always hungry), and I would come down a mountain or get to a road crossing and there would be trail magic. It never failed to lift my spirits and make me more determined to keep going.

The amount of help of total strangers was mind boggling. I remember one time I was lined up in a supermarket checkout purchasing my re-supply food, and I guess I looked like a homeless person or a tramp. The lady in front of me said to the cashier that she would pay for my items. I tried to explain that I was a hiker and could well and truly afford to pay for my supplies, but she insisted and then proceeded to ask me if I had a place to stay. I said I was hiking out that day, but it ended up with me going home with her, having a shower, my hiking clothes washed and dried, some great food, a comfy bed for the night, and a ride to the trailhead the next day. How good is that random kindness? That would not happen on an Australian hike.

So the days continued and the miles to Katahdin ever decreasing. I struggled with the idea of my limited time as my visa was for 180 days. I had wasted 40 or so days healing my ankle. Do I speed up and go really fast or do I slow down and enjoy the experience but not having the

experience of completing a thru hike? For many weeks I
slowed down and hiked with a few others enjoying the
company and setting up in a shelter early afternoon to
play games of cards, hanging around the fire chatting, and
generally goofing off. Our little group eventually
dispersed and I once again hiked alone. I did this for a
few weeks until I met some amazing young men who
were thru hiking. We formed a very unlikely bond.

We called ourselves the "A Team" and we crushed miles
and miles. They encouraged me to hike further every day
challenging me all the way. Suddenly, I was once again
inspired with the thought that I could actually finish this
hike. I would start out early, around 6:00 a.m., while they
slept. I would hike fast trying to get to the designated
camp before them. In the beginning they would catch up
and pass me before lunch. As my fitness improved and
my motivation increased with the challenge of beating
them, it was catch up after lunch. I actually managed to
get to a campsite only once before the rest of the A Team
in those amazing days of "crushing miles". We had some
great hiking days and formed a close bond, friendship,
laughter, hard hiking days, and above all, respect. I will
forever remember my teammates with great affection.
They turned my hike into the most amazing experience in
my life.

Days raced by. My time was running out. We got to the
top of Mt. Washington and I knew I had to leave my trail
family, the A Team, as well as the trail, and return to
Australia with 300 odd miles of unfinished trail. For me
it was heartbreaking. Even now as I write this many years
later I can recall my exact emotions. For some time I had
known that I would not be able to complete the trail and
that I would need to return someday to do the last few
hundred miles. I knew then that nothing would stop me
from returning and having another go. The decision was
made in my head that the next year I would do the whole

trail again just to get those last few miles in and arrive at the top of Katahdin triumphant.

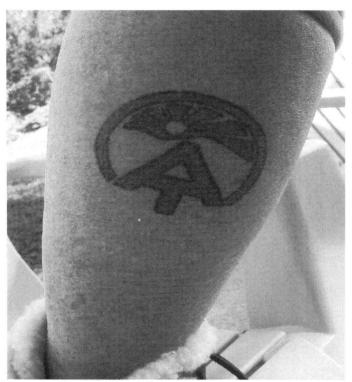

I returned to Australia not the same person that left. I had changed physically, emotionally, and mentally. I could not settle back into society and "normal" life. My preparations started right away for my return next year.

March 2011, I am back in the States beginning my thru hike of the AT. This time around I started off with a fierce determination to become a Thru Hiker! Nothing would deter me from getting to the summit of Katahdin. This hike for me was so different from the previous year. Once again, I met hikers, had trail magic, stayed at many hostels, and caught up with so many wonderful people I had met the year before.

It was a wonderful time and having the experience of the previous year made the hike less mentally and physically challenging. I summited Katahdin on the 1st of September 2011 with a feeling that mere words cannot describe. The following year, 2012, I once again returned to the States to hike some of the Pacific Coastal Trail. Then, crossed the country to the AT and completed my 2010 hike, so that I could claim being a Thru Hiker twice.

Six years have passed since first stepping onto the AT. I think of my journey often. At odd moments I have a flashback to those awesome days of hiking and know I would really like to do it again. I still hike mainly in Australia but its not the same… Maybe I have one more long distance hike in me. If I do, I know it would be the Appalachian Trail.

~ ~ ~

Long distance hiking must be in my blood. While my mother was pregnant with me my parents and older siblings escaped over the Alps from the Russian occupation of Hungary to a refugee camp in Austria where I was born. Eventually we made our way to England and from there to Australia where I grew up. I was six when I received my first pair of shoes which I promptly threw away.

I've lived in South Africa, Papau New Guinea and all around Australia where I have raised two remarkable sons. This is my story.

Chapter 4 How I Met Myself
Ryan Holt aka Yukon

September 18, 2012:

Today is my 28th birthday and I've just walked over 2,000 miles. Breathing deeply, heart thumping from my chest, the world around me seems to be disappearing as my entire life begins flashing before my eyes. How did I get here? All the events in my 27 years since birth have been leading me to this singular defining moment in time and space atop the greatest mountain in Maine, Mt. Katahdin, and the summit is only a few hundred feet away.

To better understand the depth of this journey I may need to back things up a bit.

Maine is where I was born and raised, the only State bordered by only one other state and the only state with one syllable. Aside from those uniquely fun facts, Maine is a state of purity, beauty, and wonderment for the outdoor adventurer-enthusiast.

Bordering the Southwestern Mountains, Harrison is my home town, my stomping grounds until I was old enough to venture out into the infinite abyss on my own. It's a quaint little rural town nestled between two lakes. Locals refer to it as "The Friendly Village", where everyone knows your name. People wave as they pass by and always greet one another with a smile. Our close-knit family of four lived right in the village and I can remember spending most of my summer days at Crystal Lake Park, swimming, playing kickball, and street hockey. On rainy days you could find me at the general store in town playing one of the two arcade games they had tucked away in a dusty corner. Sometimes I'd get lucky and find a spare quarter in the coin return or a nickel under the table. I would immediately raid the penny candy shelves and life was good!

It's the little things. Looking back, I see how fortunate I was to have been raised in such a wonderful setting with two loving, supportive parents, and my older brother Tim, who I always insisted on tagging along with. A sense of community, kindness, and giving are a few lessons I have carried with me from my childhood and for that I am grateful.

As a young boy, I had quite the imagination with an energy that challenged most teachers. I guess sharing, laughing, conversing, and connecting with others has always moved me. My active little aura kept me outside most days, playing into the late evening until the street lights flickered on and my folks had to yell for me to come inside for dinner. Camping, running through the forests, town tag, paintball, and building forts were a past time. At such a young age my worldly views were fairly narrow, yet "my world" seemed to be limitless in every adventure I immersed myself in. A sense of adventure intrigued me and around the age of 10 my good friend John Wielki and I made a pact to one day become United States Marines. From that moment on, we never even considered another option. Of course, I did not understand the weight of this responsibility just yet and I could not have predicted the Afghanistan and Iraq wars, but In hindsight I see that our pact was a dream for us to expand our boyhood adventurous minds and continue to explore unknown landscapes as strong, capable, exemplary men. This dream would eventually become our reality and it would all begin to happen so quickly.

At the speed of life, high school flew by in what felt like mere flashes of time. We surely made the best of those unforgettable years, if only we could remember them. ... Following graduation I had one last summer as a kid and on September 1, 2003 a greyhound bus was delivering me to Paris Island, South Carolina, where "Marines are born". My enlistment wasn't for freedom, it wasn't for God, and it wasn't for country. At 18 I was still just a kid and all I knew about life at this point was I'm about to fulfill my childhood dream and that's all there was to it. Marines are a unique breed but the one promise I made to myself before leaving home, is that I would never fall into the "brainwashed" stereotype. I understood that I was offering my life for this service and I would honorably uphold my oath to this country. But what I feared most, was losing myself, more than death. I didn't care if I was to lose all my limbs, if I could just come back from this experience with everything from the neck up, I knew I would be OK.

No one was going to take "ME" from Me and in the end, some of Me would be left standing.

Marine Corps boot camp is the toughest initial training of any branch of service and the longest at thirteen weeks. I was intrigued by the high standards that set Marines apart from the rest -- the best of the best with a sense of invincibility. It was the perfect recipe for a young man with an influential mind looking to be part of something bigger than himself. My intentions were not for the world to fear me but for people to look up to me as a balanced and well-rounded inspiration.

The moment I got off that bus in the dark of night and stepped onto the yellow footprints with a face full of spit, I knew my ass now belonged to the Corps. I learned my place quickly. There was no other way but to adapt as quickly as you possibly could and for me failure was not an option. My individual identity was stripped and I was now officially classified as, "a worthless, no good, piece of shit". They take everything away from you so fast and with such intensity that you don't even know your own name or which way is up. In a chaotic production line they herd you through like sheep to issue your uniforms, toiletries, and patchy shaved heads. You're then assigned a Company, a Platoon, a weapon, a Rack number, and a foot locker. That was everything I owned and my world as I knew it.

It was actually quite impressive to witness and to be a part of but it would take a strong mind, body, and spirit to follow this dream through to the end. The first two months felt like an eternity; there was zero end in sight. Don't get me wrong, it was new and exciting, and I had no regrets. I was following my path, but it was challenging and all of my senses were on high alert 24/7. In the beginning I had the attitude that I couldn't be broken but Drill Instructors aren't easily fooled and will make it their life mission to break you. I was put in my place more than once and grew up fast within these three months.

Every hour we were learning something new. I could feel myself getting stronger and my awareness expanding with every little new detail being repeatedly drilled into my existence. There was never a dull moment with physical training, weapons cleaning, drill formations, quarter decking, dressing by the numbers, hazing, swim qualifications, rifle range, more quarter decking, and more hazing, etc. There wasn't time to question anything or miss home, "YES SIR!", "NO SIR!", "I SIR!", that's all there was time for. One day while at the rifle range, all seventy recruits were lined up in the squad bay when Recruit Sharpe runs up to the Senior Drill Instructor's hatch and pounds on the wall asking permission to speak with him. "SIR, RECRUIT SHARPE REQUESTS PERMISSION TO SPEAK TO SENIOR DRILL INSTRUCTOR STAFF SERGEANT JOHNSON!!!" He must have slammed on the wall and repeated himself five times yet no answer. It was forbidden to do anything without permission. Recruits peed themselves quite often and I can vividly recall another recruit so scared to ask permission that he soiled himself while conducting partner assisted sit-ups -- I just happened to be the unlucky partner who was holding his feet.

This particular morning our heavy hat Drill Instructor must have been having a bad day because he flew across the squad bay in a flash, yelling, "HE'S NOT HOME!!" veins popping out of his head, spit flying through the air, and extending his knife hand at Recruit Sharpe's temple like a battering ram. Upon contact, Recruit Sharpes head bounced off the concrete wall he was standing beside and a fountain of blood spouted in a giant arc all over the cold white tile floor. You could see every sphincter in the room tighten up and all seventy recruits standing on line were wide eyed and frozen like deer in head lights. No one dared to twitch. Our D.I. retreated into the bathroom with the mess he created only to return after about 20 minutes of motionless silence and "instructed" us as to what happened:

"Recruit Sharpe was running down the squad bay to the bathroom, he then slipped on the tile floor, and slammed his head off this here corner of the table. DOES ANYONE HAVE ANY QUESTIONS!?"

"NO SIR!" we ALL replied.

With the voice of a wrathful God he reiterated his question, "If anyone is going to tell any other version of this story, be a man and tell me now!?"

"NO SIR!" we ALL yelled with a slight tremble.

My eyes were wide open now; this was a serious path I had chosen. "The Friendly Village" was fading fast from my rear view mirror and my world was expanding at an accelerated rate. I never lost sight of why I was here and all the while keeping a little piece of "ME" far away from the grips of those who now owned me. I applied myself in every direct order, test, and qualification and before I knew it the final days had arrived. November 23, 2003 was graduation day. I had fulfilled my childhood dream of pinning on the Eagle, Globe, and Anchor, earning the title of "United States Marine". It was the most surreal moment I had ever experienced up to this point in time and it was something that could NEVER be taken away from me. The first of many hats I would wear and dreams I would create into my reality.

Following boot camp I endured another nine weeks of specialized Infantry Training at Camp Geiger and was assigned to my first unit at Camp Lejune North Carolina. 3rd Battalion 6th Marines, India Company, 3rd Platoon,

1st Squad, this was my family and my home for the next four years. Less than Two months after being assigned to 3/6, my brothers and I were deployed half way around the world to fight the war in Afghanistan. At first, none of it felt real, but it all sunk in real quick upon hearing the snap of the first bullet past my head. No other reality existed.

The constant alertness, sleep deprivation, and knowing that each day could be your last takes a toll on a man. Threats from every direction, every rooftop, every vehicle, garbage piles, or animal carcasses, not knowing when your ticket might be punched and the responsibility for the lives of your brothers weighs heavy. I carried much of this weight long after my deployments had ended. I was drowning in regret and the, "What If?" game was a plague to all of us that shared in the evils of war.

I would be lying if I said I hadn't any inner demons following multiple deployments to the front lines of Iraq and Afghanistan. I soon realized the only thing we were over there fighting for was our brother to the Left and Right, nothing else mattered. It was a different kind of war; we were up against an unknown enemy who didn't wear a uniform, shooting from mountain tops, and retreating into Pakistan knowing we couldn't follow in pursuit. They would blow themselves up if they thought they could take one of us with them. It was never a fair fight and there was no value for life. I will never understand this lack of compassion for our fellow Man; life is so precious and too short to be so hateful.

To this day only a few at the top know the real reasons why we invaded the Middle East but as far as we were concerned our mission was to keep each other alive, period. Some of my brothers made the ultimate sacrifice. They gave their lives so I could live and for that reason I refuse to live a mediocre existence filled with regret. I

refuse to be selfish. I refuse to live a life that isn't
ultimately free, peaceful, and happy. I owe this to them.

That Ten year old boy who made a pact with his best
friend to become Marines may have gotten more than he
bargained for and he's now a distant childhood memory.
After Eight years I was somehow more lost and had more
questions than ever before. Good Marines don't ask
questions, they just do what they're told. So when all these
unanswered questions clouded my mind following the
war, I knew it was time for me to move on. The deepest
part of my being could sense there was a higher calling
for me in this world and the Marine Corps was just a
stepping stone, a mere chapter in this book of life with no
ending. I was ashamed for my part in such an unnecessary
war, ashamed for the reasons behind the curtain and
hidden agendas, yet I was proud, for I had served my time
honorably. I was a leader who upheld the rules &
regulations as well as my oath, but more importantly, I
never lost sight of my own moral standards of how I
conducted myself as a human being or how I treated
others along my path. I valued this more than anything.

To be proud and ashamed at the same time for a third of
my life dedicated to this experience was a hard reality to
wrap my head around. The struggle was real. I
experienced for the first time, depression, anxiety and
substance dependency as I blindly tried to find my way
out of the darkness. I had never felt more alone. My eyes
had been opened to one way of the world, a destructive
way and it left a bitter taste in my mouth. Unaware that I
would soon be introduced to another way allowing me to
see the world through new eyes, I began planning the next
chapter of my life three years prior to my discharge.

I had my sights set on the Appalachian Trail, a 2,184 mile
journey that would take me over the Appalachian
Mountain range through fourteen States beginning in

Georgia, and I was going to walk home. I researched it every day. I yearned and longed for the day I would be able to set foot on the trail and experience the RAW freedom America "believes" we all had been fighting for. I used to have the most vivid, most realistic dreams I've ever dreamed. Hiking the trail, meeting wonderful like minded people, laughing and smiling up the mountainous coast. ... Then I would wake up in my empty white walled barracks room with over two years remaining on my contract. Those dreams were nightmares; painful, they felt like a cruel joke but they were a taste of what was to come and patience for this indescribable journey ahead was a great lesson and undeniably worth it.

Following my Honorable Discharge on October 1, 2011 I wasted no time. I couldn't embark on the Appalachian Trail until early Spring and I was not in the head space to just be hunkered down for the Winter with my demons. This Country I had fought for was actually quite unfamiliar, I had never been off the East Coast so I loaded up my '83 Volkswagen Bus and once again left home. I headed west with no plan except to make the biggest circle around the country as possible, visiting as many National Parks, monuments, and odd attractions along the way.

My first stop was Niagara Falls. I had never witnessed something so powerful created by Nature. I could feel her vibrations through my entire body. This feeling gave me a rush, it filled me with something more than excitement and I wanted more. Continuing west through the Bad Lands, Mount Rushmore, Yellowstone, and Grand Tetons, the magic was incomparable to anything I'd experienced prior. I bathed in a dozen hidden hot springs as I explored Oregon and soon met the Pacific Ocean for the first time, cleansing myself in her frigged salty waters.

I slowly began to realize that in Nature I no longer felt lost, through the Red Woods, Yellowstone, Sequoias, down to Joshua Tree, and the Salton Sea, I was immersed; I felt at home. I felt alive. I connected with hundreds of wonderful people and realized we are all the same, living our lives somewhere and figuring it all out along the way. Searching to achieve peace, happiness, and freedom, or most importantly, the answer to everything we seek, LOVE. The alignment of it all was too profound to call coincidence; it was as if I was being guided by something much greater than myself. Some may call it God, but I like to think of it as "The Great Mystery". Looking back I now realize that this was actually the beginning of my major life transition. Nothing would ever be the same and I was slowly waking up.

With the Great Mystery on my side, I slowly made my way back East, exploring Mt. Zion, Painted Desert, Petrified Tree National Forest, the greatness of the Grand Canyon, and I even had the opportunity to rock climb in the Garden of the Gods while visiting a friend in Colorado Springs. My entire being was in awe! Nature has its way of putting things into perspective, our place in the vastness of the Universe and what truly matters most. I had never felt so small as when I stood in the valley of Yosemite or when tried to wrap my arms around the "General Sherman" Sequoia Tree (Largest tree in the world by volume), and at the same time feeling so infinitely expansive as I felt the interconnectedness of it all. It was a struggle to find acceptable words that could describe what was happening to me. I was often left speechless and for no reason at all grinning from ear to ear. 12,000 miles and three months later I had made my way back to Maine, where I began my final preparations for an adventure that would take all of this to a whole other level.

Spring had finally arrived and life was in bloom, a perfect time for new beginnings. Once again I hopped on a Greyhound bus and left home making my way to Amicalola Falls State Park in Georgia, the beginning of the Appalachian Trail and my greatest adventure yet. On March 27, 2012 I took my first step on the approach trail and with 2,184 miles ahead of me I was eager to create each day, learning what the trail had to teach. Those dreams I used to have in my barracks room were now reality, Sergeant Holt was long gone and Yukon (my trail name) began his journey north.

Its not the fear of the unknown that keeps people from leaving their nest, its the fear of leaving what is familiar or what is comfortable. If you have faith that all your needs will be met and you put forth pure intentions into every choice and every action, you will receive what you put out tenfold in return. I fell in love with the trail Immediately and I knew I was right where I was supposed to be, disconnecting from all the chaos to reconnect to myself and the natural world. Nothing had ever felt so right.

For the next Six months my sole purpose in life was to get up and walk, stopping whenever I felt like it or taking a day off to let my body rest. I never said "No" to any opportunity that presented itself. I'd get off trail for a five day music festival or enjoy the beach of a remote wilderness lake or mountain summit just because I could. I often spent an extra day at one of the many unique hostels along the way. This was the complete opposite from eight years of rules, regulations, and following orders. I was free for the first time in my life and I was going to embrace it with every ounce of my soul.

I met hundreds if not thousands of people, on and off trail. We were all hiking the same trail but each person was on their own path. Some major transition, be it retirement, graduation, divorce, military discharge, or midlife crisis, it led us all here to hike this trail and learn what it had to teach. A common phrase among hikers is, "The trail provides." There couldn't be more to this truth. It provides for you on every level, physical, emotional, and spiritual. It gives you many answers or at least the time, space, and clarity for you to answer your own questions. It allows you to accept what is and what isn't. It gives you the strength to surrender things of your past which no longer serve you or others.

Life flows through you on the trail. Instead of holding on to destructive forces and the extra weight of anything you might be carrying, you begin to let go, leaving behind a trail of bread crumbs and feeling lighter with each mile you hike. You begin to realize that we are not just visitors in the natural world but very much a part of it, connected to everything. This connection was something I could physically see, emotionally feel, and would soon come to spiritually KNOW.

Just as I began to experience on my road trip, every day seemed to be guided. The magic was undeniable. I found that the more present I was and the more in the moment I lived, the more attuned I was for these profound alignments to happen on a daily basis. I was literally creating each day with my thoughts and intentions, manifesting my needs and even some of my wants. I had never felt so in control of my life but the funny thing is I had to give up all control in order to achieve this.

On the morning of July 4th I crawled out of my hammock and joined a few of my brothers around the camp fire for breakfast. Boomer, Squatch, Pops, Daffy, and myself began our day just as any other, except this particular holiday morning Pops boisterously asked, as we ate our hot oatmeal and pop tarts, "Well boys, what do we want to create today? Its the 4th of July. What do we want to manifest, what do we want to see happen on the trail today?"

Clockwise around the circle we all verbalized our intentions and shared what we wanted to create. Squatch spoke first, "I'd like to find a nice swimming hole, where we can get some sun, relax, and clean up a bit." Boomer said, "I'd like a ice cold beer and a bacon double cheese burger." Daffy added, "I'd like to come across some trail magic." Pops said, "I'd like to cross paths with some familiar faces, fellow hikers we haven't seen in awhile." And, in an almost jokingly fashion, I finally added, "Well, it is the 4th of July, I'd love to watch the movie The Patriot, what could be more fitting than that?"

With all our intentions in the center, we finished our breakfast, packed up camp, and hiked onward. We had been hiking for a few hours when one by one we strolled down a hillside and were greeted with a colorful hand painted sign that read, "Trail Magic 0.2 miles ahead". Signs like this tend to release endorphins in every hiker as they turn on their boosters to see what goodies and sugary beverages await them. We could smell smoke from the grill and hear waves of laughter before we even made it out of the tree line. Sure enough, as a road crossing and dirt parking lot came into view, there was a gentleman with his mini van opened up like a food truck, offering hot dogs, chips, and coolers of ice cold soda. As we approached closer, we recognized a few familiar faces in the small crowd too! Mountain Spice, Earthworm, and Yodeler!! A few of us had been hiking with them on and off for the better part of our first thousand miles and Yodeler was a good Marine brother of mine with whom I served my last three years in Marine Corps.

What were the odds? A few minutes later and we would have completely missed each other. They were waiting on a ride from Earthworm's parents to celebrate the holiday weekend. We had a short visit, caught up on trail gossip, and once again went our separate ways leap frogging up the trail. The day was still young and Daffy had already created our trail magic and Pops had created our reunion with friends we hadn't seen in months.

"It's just coincidence," you might say. Or, "It could have happened to anyone." This is true, but then again the day wasn't over. Shortly after our reunion and trail magic we came to an old iron bridge that was now used as a bike path. The bridge was a historical work of art, but it was the river under the bridge that caught our attention. It was picture perfect July 4th weather and the river was lined with hundreds of families enjoying their freedom, fireworks, BBQs, and endless fountains of cold beer.

We had a choice to either tackle the next 3,000 feet climb out of the gap or take advantage of this beautiful day and flowing free stone river. The choice was an easy one for me. We bushwhacked through thicket and found a more secluded and peaceful bend on the river bank. What a welcoming sight! The Sun was directly overhead of us in a cloudless blue sky. My whole being was thirsty and I was excited to quench my soul in this river. My pack was still dragging on my arm as I ran into sparkling clear waters. After nearly a week without a bath the rushing water over my body was liberating.

Squatch had created this swimming hole for us and we fully embraced it with pure joy and three hours of playing, swimming, and napping like children. As early evening approached we had to make another decision, make our way north on the trail and find a place to set up camp for the night, or hitchhike to the nearest town to continue celebrating this fine day. After some indecisive banter within the group I decided to bushwhack back up the hill and stand where the trail continued north. At the dirt road I noticed an SUV parked at the trail head with MAINE license plates.

Now, people from Maine are all basically neighbors and we get very excited when we run into another Mainer while traveling outside the state. We always think to ourselves, "Maybe I know them." Overly excited, I introduced myself to the owner of the car and, as any Mainer would expect, we hit it off like old friends. She was waiting for her son to come down off the trail. I began asking questions about what amenities were near by if we wanted to get off trail for the night. She was very friendly and extremely helpful with information about the local area and even suggested a KOA campground that was eight to ten miles down the road. Not an impossible hitch but that's a decent hitch hike for five of us at this time of day. We were discussing logistics and who would hitch hike with whom when my new friends from Maine generously offered, "We'll have to make two trips, but I'll give you all a ride."

If you allow it, this is the kindness on the trail that happens on a daily basis. Everyone I met was willing to help another person in any way that they could. It was contagious, it made you want to do more for others and give back whenever possible. Once at the KOA campground we broke in to our muscle memory routine of setting up camp, changing out of our hiker funk and cleaning ourselves up a bit before venturing down to the main lodge. The lodge was a beautiful timber framed structure, post and beam cathedral ceilings, with a massive fireplace, lounge and bar area. For the most part we had the place to ourselves and wasted no time claiming the first five bar stools front and center, grazing their menu all the way through while our mouths salivated.

Boomer barely glanced at the menu. I don't think there was any question as to what he was having as he had created this alignment around our morning camp fire. "Bacon double cheese burger, French fries, and a pint of your finest suds." "I'll have what he's having," Daffy said. "Better make that five, all the way around," Pops interjected.

We were on cloud nine, all of our needs met and here we were satisfying the thirst of our hiker hunger while tipping back a few well deserved beers. Lost in my own thoughts for a moment, I was sipping on a cold brown lager while glancing around the lodge, when all of a sudden I get checked by an elbow that nearly sent me off my bar stool. It was Pops. I looked at him like, "What in the hell was that for?" He did not speak yet his eyes were wide, mouth open and pointing vigorously up at the half a dozen flat screen TVs mounted up behind the bar. *The Patriot* was on every TV screen! We all erupted in a bit of disbelief. I was speechless.

Magic like this happens on the trail daily but this profound alignment was almost perfectly executed in the realm of what we are capable of creating when we are attuned. There truly is no such thing as coincidence. We had manifested the entire day with our thoughts and intentions.

Manifestations like today would begin to happen more frequently as I moved forward with this new understanding. I would never question it again. Joseph Campbell once wrote, "People say that what we are all seeking is a meaning for life. I think that what we are seeking is an experience of being alive." His words have always resonated with me. I couldn't imagine not feeling alive; I would rather be dead.

Sometimes I live an entire lifetime in a single day. The less I have the more wealthy I feel. On that day, we were the richest men in the world. The trail does provide and it would continue to provide countless beautiful happenings and alignments. When I crossed over the last State line of New Hampshire into my home State, Maine, I could feel how much I had changed and had grown on every level since I began this journey over five months ago.

They say, "The last one to Katahdin wins." Meaning the person with the longest hike, the longest experience, is the winner. I wasn't ready for it to be over. There was no plan for me after this and I didn't want to come down from the levels of happiness and freedom I had achieved. Little did I know the proverbial "Trail", trail of life, would never end and I could bring this new knowledge along with me into the next chapter. I should have known better than to question the path in which I was being guided. There would be so much more to come, but, Mt. Katahdin would be my final teacher along this part of the journey.

September 18th 2012:

So, where was I. ... Oh yes, I'm almost at the summit of the greatest mountain in Maine and my entire life is catching up to the present moment. It hasn't even been a full year from my discharge date, yet with all that I have experienced in this short amount of time it feels like a decade has passed. Since, I've traveled 12,000 miles around the country through more than two dozen states, visiting all National Parks and Monuments and hiked for six months over 2,000 miles up the East Coast through fourteen states.

My goal to hike the entire length of the Appalachian Trail completely exceeded my dreams and expectations. The connections I discovered within other beings and in the natural world was beyond the comprehension of my younger self. This chapter was about to be fulfilled in the next few hundred feet. ... My heart pounding, breathing deeply, I can only focus on the summit sign. Moving forward, my body feels like it's vibrating immensely and the rest of the world seems to be slowly disappearing, almost like tunnel vision I can't even see my feet. I'm stumbling, kicking rocks, and a maverick wave of emotions is building rapidly with every step towards the sign.

You know how they say your entire life flashes in front of your eyes right before you die? Well, I believe the same to be true right before you wake up. It's all happening so fast. I see my family, my childhood, high school years, then memories of fighting the war and all my life's obstacles are flashing towards me like a freight train. The beautiful landscapes of this country and human connections I formed over the last year flowing through me and the 2,000 mile journey I am about to conquer trailing behind. These are the most powerful moments I have ever experienced. The summit sign is now within arms reach, a combination of every single emotion one can feel is hitting me in the face as I reach out to latch onto the sign with both hands. A breaking point, just in time as I nearly collapse from this profound happening. Time froze as I lay here hanging on, tears of sadness and tears of happiness trickling through my beard before splashing onto the rocks below. ... I was in this state of limbo for what seemed like an eternity, and then ... I woke up. Complete stillness and silence, a very heavy weight seemed to be lifted from within. I had let it all go. I am awake for the first time in my life, towering atop the greatest Mountain in Maine and reborn 28 years later to the day. With a glorious smile upon my face I tilt my head back laughing boisterously at the sky.

Ultimate peace, happiness, freedom, and the clearest state of consciousness consume my being as I sit here admiring all of creation through new eyes with a Man I finally had the pleasure of meeting. ME.

~ ~ ~

A few short months after completing the Appalachian Trail I purchased 42 acres in the Southwestern Maine Mountains with the intention of creating a Nature retreat hostel for future thru hikers and outdoor enthusiasts. This was a vision that came to me while I was immersed on the trail. It was a long term dream that I would never lose sight of. I spent the following Four years traveling and exploring more of this country while maintaining what I had attained on the trail in terms of getting to know more about myself and this newfound enlightenment, defining this new direction, my purpose and place.

The omens were abundant and listening to guiding forces not once led me astray, leading this journey full circle once again to the present day. Another dream created into my reality: "The Human-Nature Hostel" will be open for hiker services this coming Spring 2017.

I've surrendered to the forest and dedicated myself to healing, teaching, and inspiring others through the powerful medicine of Nature. There is no greater purpose than providing a service to others. At The Human-Nature Hostel and as a Registered Maine Guide I dedicate myself to a life of service to all of you and as a steward of the wilderness. I'll see you down the trail and always remember to Live-IN the Dream.

www.Humannaturehostel.com

Chapter 5 There Are Bears Out There
Danielle Breakell aka Red Panda

Unfortunately, I must admit that I was about as interesting as a stump when I was a kid, totally content to live vicariously through others. Summer vacations were spent camped out on my bed, buried under a pile of books. I much preferred to read than to go for a hike, as my father is all too happy to remind me. There was nothing better than forgetting about my life and losing myself in someone else's story. Have you ever read a really, *really* good book, and after turning the last page, you wistfully wonder, "Why can't my life be like that?", and then spend the rest of the afternoon daydreaming about mystery or magic or true love?

I am a sucker for adventure. I loved delving into someone else's head as they fought their way to freedom, escaped almost certain death, and triumphed over evil. Sounds fun, right? Who doesn't want their own happy ending? And an amazing story to tell about how they got it?

Luckily I eventually wised up and realized that embarking upon my own real-life adventure might just be as entertaining as reading <u>Harry Potter</u>.

Let's take the Appalachian Trail, for example. Going on a thru-hike sounds awesome, right? A lot of people have never considered it before. Hiking an ENTIRE trail. A trail that crosses through 14 states. It's a huge undertaking. Why would you even think of it?

But then you meet an acquaintance, perhaps someone in front of you in the line at the pharmacy, or a grubby, stinky guy with a backpack wandering through the supermarket with a desperate intensity. Somehow, you learn their story. Maybe someone else in line asks about the "2000-MILER" patch on their jacket, or you overhear their conversation with another grubby hiker. It doesn't matter how… but you hear about their journey.

Maybe they talk about the view from the top of Spy Rock at sunset, the tranquility of living separate from the stress and bustle of society, or the peace that comes from the absence of the hum of electricity at night. Maybe they tell you about the woman who hiked with a prosthetic leg, or the funny guy from Germany who only had ONE set of clothes rather than the luxuriously more common two ("You think I smell bad? You should have met the guy who I slept next to in the shelter last night!"). Perhaps they mention the woman in Massachusetts who let them camp for free on their lawn AND gave them homemade cookies, or the huge free "trail magic" picnic that they stumbled into last weekend.

Or maybe they pass on nothing more than, "Yeah, I hiked the whole thing last summer. It was the best thing I have ever done. If you ever get the opportunity, you should do it."

Somehow, you learn about their adventure. It sounds awesome, doesn't it? To sleep under the stars every night? To receive random acts of kindness from total strangers on an everyday basis? To live free from the worries of paying for rent, filling up the gas tank, and completing that report on time?

And the daydreaming starts to set in. At least, for me it did. Alexa, my park ranger friend, passed on a few stories of her trail adventures, showed me a few pictures... and I was hooked.

She had hiked and camped with a friend she met on the trail, sleeping on barely more than a tarp in the open summer air. She had taken on a trail name, given to her by someone else, just like her legal name, but with a different significance. She had summited mountains, viewed incredible vistas, and made the journey of a lifetime. She had stayed in crappy motels, luxuriating in hot showers after days of being grubby. She had hiked for *thousands* of miles, becoming an incredible type of athlete.

Alexa spoke of her time on the trail with such nostalgic passion that I couldn't help but feel jealous. The greatest adventures I had experienced weren't my one, only those written by other people. I started to think to myself, "I could do that, couldn't I?" which evolved into, "Maybe I should do it." which next became, "I wonder if I could afford it." and then, "Who cares what it takes, I AM GOING."

Adventure is so tempting. But then you start to think about it some more…

Living in the woods… where do you sleep? Where do you go to the bathroom? How do you bathe? Where do you wash your clothes? HOW DO YOU ACQUIRE PIZZA? This last one was a serious issue for me, as I have a serious addiction to pizza. Have you ever seen a Pizza Hut in the forest?

And THEN you remember… there are BEARS in the woods. Bears, and ticks, and spiders, and poisonous snakes. Not to mention creepy crawlies, mice, coyotes, foxes... and don't forget rabid raccoons, foaming at the mouth in mid-afternoon summer heat, ready to sink their claws into your meaty human flesh.

Not to mention the people. Crazed psychos, hiding from the law, prowl about the woods to avoid the police and certain imprisonment. Crazed lumberjacks, like the kind in horror movies, are just waiting for you to wander past their dilapidated cabin so they can feast on your flesh.

It's a scary world out there. TERRIFYING. And that's when you start second-guessing yourself. Maybe you don't want adventure after all. Maybe it's much more sensible to just sit in your cozy little home, snug under your covers, and live vicariously through a fictional character.

I admit that in my personal case, I kind of neglected to fully take all of these factors into account. Bears? Pfft. I'll hang my food from a tree. Ticks? I'll slather on some DEET. Psychos? If you haven't noticed, hiking poles are fantastically sharp. Who needs a bayonet when you have a pole?

That's not to say that I didn't consider those dangers. Of course I worried about bears. I worried about getting stalked by strangers. I worried about Lyme disease and getting lost and falling off of a cliff.

I was hesitant to tell my parents about the journey I was considering. Would they take me seriously at all? It wasn't like I had any backpacking experience. They would just think I was crazy. Surely they would say, "Maybe, honey…" over the phone then roll their eyes and think to themselves after hanging up, "It'll never happen."

The thing is, your friends and family love you. They don't want you to get eaten by a bear. They don't want you to meet an early, tragic demise by careening down a mountainside, thumping your way over boulders until a nice, thick tree trunk stops you dead.

And if you just happen to be crazy enough to commit, or to tell a select few friends and family members your harebrained scheme just to test out the idea, you will quickly learn that they will be more than happy to point out all of the disastrous things that might happen to you.

"Don't you think you'll run into a bear?"

"You have to be really careful about Lyme disease. There are ticks everywhere out there."

"What if you run out of water?"

"Didn't I just read about someone who got lost out there?"

"But you might not have phone service in the woods."

"YOU ARE NOT ALLOWED TO HITCHHIKE." (My mother's boyfriend was very emphatic on this point. I didn't listen.)

All of this, as if you are a full-blown idiot who hasn't done ANY research into the theory of spending FIVE MONTHS in the woods.

After a while, you get sick of people telling you these things. You know from the moment concern clouds their features and they start to open their mouth that you just have to suck it up and nod and pretend to care.

"Are you going to take a gun with you?" They will inevitably ask (I even had one aunt bring out an example of a nice, lightweight handgun as an example), and you resist the urge to smack them, hoping it might knock some sense into their brain. You resist the urge to reply, "Yes, definitely, I am going to bring a GUN with me, and keep it in my backpack, despite the fact that the laws differ in each state, I have never taken a gun safety class, and I would be practically inviting any crazy person who might rummage through my pack while I'm sleeping to wield a deadly weapon.

PLUS, DO YOU KNOW HOW MUCH A GUN WEIGHS? You have to carry that sucker with you ALL THE TIME. Hikers chop of the bottom halves of their toothbrushes to save weight, jamming little stubs back and forth between their lips to clean the remnants of their nasty hiker diet off of their teeth, and you expect me to carry something that will take up space and weight and has the potential to KILL someone, including myself?

No thanks, dudes. Not for me. I'll take the risk of sticking with my hiking poles. Those things can really be deadly. Have you seen the points on them? If you get some good momentum going, they could cause some serious damage. I know partly from experience because one time I accidentally slammed one down onto my boot instead of into the dirt (I never said I was coordinated... or intelligent, okay?).

That's how I felt when I first decided to attempt a thru-hike. I was so excited about the prospect of being on my own, free from worrying about finding a job and becoming a "real adult" that none of those little issues bothered me.

Of course I realized that the trail would be dangerous. And of course I realized that hiking alone brought on several potential issues. I was definitely worried about being a short, small, fairly young woman on the trail. There are many more men than women on the trail, and, well… people are human. You're out in the woods, sweaty and hot, in good physical condition, surrounded by similar people who are out there because of a similar desire to escape into the freedom of the forest. Animal instincts kick in. It's only natural that horny guys will approach solo girls in hopes of something more than friendship.

Of course, if you are a thru-hiker, you are also sweaty and gross, grubby and smelly, isolated from sanitary conditions. There are couples on the trail. Sex must happen... it's only natural. My thoughts on this matter? "Ugh, so *gross*." I feel like I could barely keep myself clean, I don't want to think about the added disgustingness of intercourse after a day of hiking through mud.

Part of my reason to hike was to help get over a long relationship, and I had ZERO (okay, maybe just a tad more than zero… we'll go with 2% for the sake of argument) interest in finding a dude to sleep with. Or kiss. Or show any interest in at all, actually. I didn't want ANYTHING to do with romance. I wanted to be free, to be myself, to be my own person and to grow independent and be happy with it. Thus I set off totally willing to tell every guy who looked at me funny that I was not at all interested, thank you very much.

Not that that would stop interested parties from coming after me, necessarily. I had been training a lot in preparation, however, and figured that if worse came to worst, I could at least run away. I had pepper spray and swift legs.

IN SUMMARY: I did NOT set out to with any romantic desires. The idea actually kind of grossed me out. And for the record, I still have only kissed two boys in my life. Both of those boyfriends were pre-trail. So there. One mission accomplished. My trail virginity is intact.

So clearly I also did not get raped. I definitely had some guys show their interest, but didn't return it. There were some close calls with overly affectionate men, but I managed to escape (when in doubt, hike faster!).

To say that I wasn't interested in a relationship of any kind doesn't mean that I didn't want friendship. I honestly figured that I would end up finding a friend to hike with, somehow, eventually on the trail, just like Alexa (my friend who got me obsessed with the idea in the first place) had. The trail is a community, and, like I said, it's full of like-minded people who are out there for many of the same reasons you are. Thru-hikers share a common goal and a common bond. I didn't think it would be hard to find a friend. And although I didn't spend a lot of time hiking alongside many of them, I did find more than several friends.

So I have admitted that at the time I committed to my hike, people were all too eager to point out the potential dangers. Blinded by the prospect of living my own adventure, I was brash and reckless and probably not as scared as I should have been? Was my cavalier regard for my own safety justified? Here's the truth...

In my five months on the trail, I never felt like I was in any SERIOUS danger. Walking through the streets of

Hartford is more unnerving than traversing five hundred miles alone in the wilderness, where rabid coyotes might be lurking behind the nearest tree, ready to sink their teeth into a hiker's leg (muscular and extra delicious!).

Was this stupid of me? I didn't think so at the time. In retrospect, I sometimes wonder what the heck is wrong with me. I was out in the woods. ALONE. WITH BEARS AND CRAZY PEOPLE. TERRIFYING, RIGHT?!

Nah. The thing is, when you're on the trail, you meld into that wonderful like-minded community. The trail is like a river. The ebb and flow of hikers around you is constant. Some days, it's a deluge. Other days, it's a trickle. But even when the current dwindles down so the river is barely more than mud, the riverbed is still there. You know that there might be a puddle of water around any corner. More water will always come.

Such is the trail. At any given time, you know that a flood of hikers surrounds you. Maybe you don't see them right now, but they're there. And after a week or two, they know your name. In most cases (unless you are beastly fast or horribly slow, there are people ahead of you and people behind you.

Those people learn your name. They look out for you. They ask other hikers if they've seen you. They pass on your stories, look for your name in the shelter registers, and wonder where you are if they haven't seen you for a long time.

Even on the nights when I camped alone, unaware as to the location of the nearest human being, I never felt alone. Of course, there could have been a bear hiding behind the privy ready to maul me, but I was blissfully ignorant, caught up in my own adventure.

Best of all? Pretty much everyone on the trail is there for the same reason you are. To escape "real" life, to enjoy time with nature. To be OUTSIDE, with the air and the dirt and the water. They want to sweat and scale mountains and tumble into their sleeping bag into a state of exhaustion, with that sweet smile of accomplishment on their face.

The journey is more than miles.

So, perhaps incredibly, ridiculously naively, I wasn't scared. I didn't feel unsafe, for the most part.

This does NOT mean, however, that I was always safe. I had several sketchy encounters. There were several times that I found myself in a situation and thought, "HOW AM I THIS MUCH OF AN IDIOT?" or "IF THIS IS HOW I DIE, I WILL BE SO PISSED."

So, in my case, there were a few specific forces that caused more misery and trouble than others. Thru-hiking never really seemed dangerous. There were a few aspects, however, that were downright unpleasant and made me unreasonably miserable. Take, for example...

Shoes and Socks

Were you expecting me to say bears? BEARS DIDN'T BOTHER ME. I never had a problem with a bear. A few ran away, a few munched berries nearby, a few looked up and stared at me in boredom for ten seconds as I shrilled on my whistle and obnoxiously banged my trekking poles together to scare them away before going right back to what they had been doing.

What's worse than a fricking bear? HAMBURGER FEET. SWOLLEN RED SAUSAGE TOES. Imagine taking off your shoes and finding a hot dog where your pinkie used to be.

Have you ever worn an ill-fitting pair of shoes all day? One that pinches your toes or rubs your heels? Kind of sucks, right?

NOT AS BAD AS HIKING TWENTY MILES IN THOSE SHOES. FOR SIX DAYS IN A ROW.

One time, I hiked 100 miles in 4 days. It was a bad choice. A very bad, stupid choice (have you noticed a pattern yet? I was a bit of an idiot of a hiker…). Now I can say that I did it, though. AND I SURVIVED.

The shoes I stared the trail with WRECKED my feet. I had huge blisters. My toes went numb. I couldn't sleep at night because of the throb, throb, throbbing. I called my mother in tears and asked her to overnight my emergency pair (a half-size larger) to the next town.

Those shoes sucked, too. I plastered my heels in duct tape to make sure they didn't rub raw. I wore two pairs of socks to keep my toes from slipping and slamming. My feet really hurt. The socks didn't help.

Speaking of socks, worse than the fear of a bear ripping through my tent was the horrible depression of accidentally slipping into a puddle. Or getting stuck in a drizzle, and ending up with...

…WET SOCKS.

A thru-hiker has to carry everything they need with them at all times. Your food, your shelter, your clothing... everything gets stashed in your pack. You can't carry an infinite amount of stuff. I only had three pairs of socks. I only had the opportunity to wash my clothes about once a week.

Inevitably, your socks get wet. You accidentally dip your foot in a puddle. You lost the genetic lottery and are cursed with overactive sweat glands. You get caught in a downpour. And your nice wool hiking socks are soaked. Thus you are doomed to a day with wet socks.

Those suckers suck. Wet socks equal wet feet. You know how your skin gets all pruney in the bath? Imagine your entire foot being pruney. All day. And imagine hiking with your feet all pruney all day, nice soft skin rubbing raw against the inside of your boots. The duct tape you spent ten minutes carefully applying that morning comes RIGHT off, and by the end of the day the skin on your toes seems to have reformed in a strangely square shape.

I DESPISED wet socks. Even if they miraculously dried on a rock while I ate lunch, they were usually so caked with mud that they took on the consistency of cardboard. It's difficult to cram your toes back into stiff socks.

Waking up in the morning and having to dress myself in yesterday's cold, smelly, perpetually-damp hiking clothes was bad enough, but I actually dreaded the mornings when I had to slide my poor sausage toes back into wet socks.

I would hoard fresh pairs in the bottom of my clothes bag so that I was guaranteed to have at least one day of fresh, clean, comfortable feet a week. Of course, even that didn't often work, because damp boots will dampen dry socks right up again anyway. Sad times. I would have much rather had a clean pair of socks every day than a hunky trail suitor.

So yes, despite several people's inclination to rank bears on the top of the list of trail dangers, I'd place wet shoes and socks up there. Good shoes and dry socks make a world of difference.

There you have it. Socks are like hiker gold. Necessary but often horrific. Want to learn about another potentially horrific but often necessary aspect of the trail? Ever think about climbing into a car with a stranger?

Hitchhiking

Okay, so you probably suspected this topic was coming. But still… in my experience, hitchhiking is surprisingly safe. The thing is, people who live in/near trail towns are used to grubby hikers stumbling out of the woods and popping up their thumbs on the roadside. They get to know the stinky crowd.

Some of the most incredibly amazing people I met on my journey stopped on the side of the road to offer me a ride. Many times, I didn't even have my thumb out. You think being a young girl on the trail is dangerous? If anything, it encourages more people to keep an eye out for you.

For example, my experience in Vermont: after pulling over to offer me a ride, a woman explained that she has a very dirty sense of humor and likes to play dirty songs for the hikers she picks up.

Did I mention that this lady appeared to be in her mid-seventies, and was on her way to church? Oh, and that *after* church, she was going to visit her husband, who had a stroke five years ago, in an assisted living facility?

She was wearing rubber chicken earrings, and as she drove me to the hardware store to purchase duct tape, my ears were treated to an Elvis impersonator's serenade regarding his desire to be exhumed so that he could beat up Michael Jackson. Also, "Mrs. Claus Cut Off Santa's Schlong' (to the tune of "Santa Claus is Coming to Town." And next? Another schlong song, but about Bill Clinton.

Upon our arrival back at the trailhead, she played the last song, which she traditionally plays for every hiker she says goodbye to. It was about boobs. To the tune of "We Didn't Start The Fire" ("Here's a little ditty 'cause we love your titties.")

Now, I really do not have a perverted sense of humor, but this lady was hilarious. Totally not what you would expect by looking at her.

Several of my hitchhiking encounters were, though perhaps not as funny, just as safe. Except for that one time that I desperately accepted a hitch into Fontana Village....

I naively thought that it wouldn't be a horrible idea to WALK into town (despite being less than 3 miles, it was a terrible choice to attempt a hot, uphill, paved journey on already-trashed feet). I was desperate to get there because my mom (aka trail postmaster) had shipped a new pair of boots there, and my feet were suffering.

After half an hour, a truck pulled over to ask where I was headed. Two kindly looking middle-aged men, reminiscent of an uncle, sat in the front seat and offered me a ride. I was so relieved and excited of the prospect of not having to walk another uphill paved mile that I didn't hesitate as much as I should have before heaving my pack into the bed of the truck and climbing in after it.

As I leaned through the cab window to thank my rescuers, I spotted the cup holder. Nestled inside was a nice, cold, beer with the pop topped. WHOOPS. Maybe I hadn't made the best choice to accept that ride after all. It soon became apparent that the nice, uncle-like driver was a tiny bit tipsy.

By some stroke of fate, we arrived at the hotel where I needed to pick up my mail without running off the road. Then, kindhearted men that they were, they offered to buy me lunch. I had to go inside to get my mail anyway. I had no escape. I *needed* my new boots, so I reluctantly accepted.

Extra luckily, there were hiker friends already there who were more than willing to relocate into the dining room where my tipsy acquaintances offered to buy me lunch. My friends were gracious enough to accept my admission, "Okay, I'm an idiot." and to keep an eye on my boisterous company (I told you… a community!). It soon became apparent that not only were they drinking, they were drunk. And quite probably alcoholic.

The thing is, they were incredibly sweet, well-intentioned men. They didn't try to come on to me, they didn't ask for anything in return for their generosity. Apart from their severe drunkenness, they reminded me of my father in their kindness and eagerness to provide some small kindness. I am glad that I met them. Not glad of the choice I made by getting in that truck, but you can't win them all. The Alcoholics, as I fondly remember them, were not safe, but still endearing.

Yes, hitchhiking is dangerous. It also, however, restores your faith in humanity when a middle-aged couple with their giant friendly dog invites you into their car, telling you that they wouldn't allow their daughter to pick up hitchhikers... unless they're *hiker* hitchhikers.

What's even more potentially dangerous than hopping into an alcoholic stranger's truck?

Weather

Everyone freaks out about the possibility of Lyme disease, but I found three ticks on my body during my five-month thru-hike. Weather, on the other hand, came close to killing me way more than three times. THANKS, MOTHER NATURE. You would think mothers are supposed to be gentle and supportive. Mother Nature is the kind of mother that kicks you out on the curb and tells you to suck it up and deal, or else you'll never learn. It's CHARACTER BUILDING.

I kind of hate summer. Okay, I love watermelon and corn on the cob and Dairy Queen and bike rides and swimming in the lake, and kayaking and all of those nice fun summer things, but I HATE the heat. One of the nice things about the trail is that you're usually, at least, in the shade.

Summer heat is an incredible force. The human body is not made to spend weeks upon weeks of nonstop strenuous hiking in that heat. On a thru-hike, you might be *climbing mountains* in the blazing sun on a 100° day.

One memorable afternoon, I pitched my tent on a nice, flat bed of dried leaves under the trees and then embarked on a journey to the stream to wash my face and change into my dry sleeping clothes. I then had the immeasurable pleasure of draping my hiking clothes, sodden with sweat, over a nearby tree limb to dry. My shirt was caked with the salt that had oozed out of my pores throughout the day.

I clearly remember wringing out my disgusting underpants, so soaked with sweat that I was amazed that the material could even HOLD that much. I tried to estimate just how many ounces of disgusting sweat they contained as I thought to myself, "What has my life become?!"

Those were seriously the nastiest underpants I have ever seen, let alone had the pleasure of handling. I was that sweaty. It was that hot. That kind of heat can kill you.

Another day not too long after that, I found myself stumbling weakly up the trail, unable to find the strength to get uphill, fumbling like a rubber puppet. It was a weird sensation. Hello, dehydration. Whoops.

After that, I started drinking more salt and sports drinks to replenish my electrolytes (if you are saying, "Duh, you should have started doing that the whole time," I just want it to be known that I AM NOT AN ATHLETE).

Anyway, the heat kicked my butt. I kicked it right back, but it was a clumsy, staggering, idiotic kick. My skills are limited.

Another fun afternoon, I got caught in a thunderstorm. After a pleasant morning, I decided to ignore the ominous forecast and keep on hiking. Soon, the wind was raging. The trees were swaying dangerously, and branches were tumbling through the air. Thunder rumbled in the distance.

It sounds weird, but I hiked with an umbrella, and boy was I glad I had that thing that afternoon. I kept it out in the wind, even though it wasn't raining yet, yielding it above me like a shield, hoping it might be semi-effective at deflecting any rogue branch that might come cracking down on my head.

Soon after, I was caught in a violent downpour. Lightning flashed and thunder cracked right after. Raindrops pelted at me sideways. The trail turned into a river as rainwater gushed downhill, soaking my boots. I felt like Dorothy in the twister, trying to keep my umbrella from blowing inside-out, trying not to slip on the smooth, wet rocks, trying to see the white blazes through the rain. With a huge sigh of relief, I spotted the stone pavilion atop Sunrise Mountain, where I took refuge until the storm lulled.

Almost dying by tree-impalement was bad enough, but I STILL HAD TO SUFFER with wet socks for the rest of the stupid day.

Good story, right?

Then there was the day I camped at Pinkham Notch. I am not totally sure why I was in such a hurry to get there. I think it was a combination of "There's nowhere closer to camp, might as well go all the way," and "If those guys are still planning on making it there tonight, it's not too late for me to go, too," and "My mail drop is there, and there are almost definitely some M&Ms in that package…" and, of course, the prospect of a hot shower (the Pinkham Notch Visitor Center has coin-op showers).

The day started fine. The day's hike began as usual. It was actually a momentous day, as it took me up Mount Washington. It was gorgeous. The sky was blue, the sun was out, and the climb was nowhere near as difficult as I had been expecting.

Of course, you may know that Mount Washington is unpredictable. They even put a nice sign in the middle of the trail warning you that "THE AREA AHEAD HAS THE WORST WEATHER IN AMERICA". So even though my climb to the summit was cheery and wonderful, I wasn't surprised as the sky started to cloud over as I hiked on to Madison Hut, which is like a large cabin with bunkrooms and a kitchen run by the Appalachian Mountain Club.

And that's when the clouds came.

I looked behind me and saw gray, lots of gray. The storm clouds were being pushed over the mountain behind me by the wind. The effect was very cool. Gray plumes billowing forth, streaming across the sky above me.

Then the wind picked up. And thunder started to rumble. And the wind tugged insistently, pushing and pulling me in the wrong direction. The sky grew more and more ominous.

I picked up speed, darting over rocks, grateful that the section of trail wasn't too rough, but more like a messy cobblestone road. I didn't have to think too hard about where to place my next step.

As I quickened my pace, it started *hailing*. Fantastic. Nothing like walking across an exposed ridge line, totally exposed, as a storm rolls in. I admit, I was a little scared. I didn't really want to get fried to death by lightning, and I was an excellent target for rampant electricity. Just imagine, a fried hiker tumbling her way down the Appalachian Trail, smoke rising out of her braided head and charred black flesh leaving a flaky path to her dead body.

You would think this would be a good sign to stop hiking. But I'm an idiot, remember? A group of familiar, friendly thru-hikers I sat next to in the Hut planned to continue on, so I figured I might as well go, too. And I did.

Thus began my next near-death experience. I departed the Hut and I followed my hiker companions back north on the trail, heading into the ascent up Mount Madison. It was a hands-and-knees kind of climb. I could not find a good way. It was *hard*, and I kept overbalancing and falling.

And that was only the freaking beginning. As I climbed up higher, it got even MORE terrifying. Hooray for adventure!

It was steep. And rocky. There was no clearly-defined trail. I tried to make my way from cairn to cairn as best I could, regardless of whatever non-existent path I was supposed to follow. The guys were disappearing into the distance in front of me. "Don't leave me here to die!" my inner voice screamed as they grew more distant and I struggled to catch up, my trekking poles skittered into gaps between rocks.

The wind was intense. It buffeted me, constantly blew me off course, smashed me sideways into rough rocks, tripped me up and pushed me over. I tried to keep rocks between myself and the gusts as a barrier, but this strategy didn't serve me too well.

If the rough trail and wind weren't enough, sporadic bouts of rain pelted down throughout this ordeal. Wet and scratched, and frustrated, I was exhilarated and thrilled but also kind of pissed off at that freaking wind. It was *so* strong, I couldn't control my own body, my own hike. It was a good thing that the guys were nearby. Their whoops and hollers of exhilaration, frustration, and amazement lifted my spirits.

The climb along the ridge seemed to last forever. I kept hoping to see the trail ahead descending back below the tree line, but after every peak, I just found another rocky stretch in front of me.

Eventually, nearly cheering with relief, I passed back down below the tree line. As I picked my way amongst the rocks, attempting not to slip and split open my skull, the guys ahead of me were flinging out every curse and swear I was inwardly thinking as we slipped and skidded our way down the slick, steep, wall of a descent back down towards civilization. The roots were insanely unsafe. The rocks were slippery from the rain, and I tripped and cursed more than a few times.

Weather, in its many incarnations, is a frequent death trap. Think you might get eaten by a bear? Maybe you should be worried about a branch through the brain, or getting struck by lightning, or freezing to death camping on Roan Mountain. The shelter there is the highest on the entire AT, and the night I stayed there hikers were crowded into the upper floor like frosty sardines, hoping to benefit from what little heat there was trapped beneath the roof.

Anyway, next time you're worried about a bear, make sure you remember to pack plenty of water and extra layers, just in case Mother Nature decides she wants to take a swing at you.

So what do I say to dangers of the Appalachian Trail? To bringing a gun, pepper spray, and an emergency GPS locator?

I was really freaking lucky that I never would have needed any of these things. Even if I did, it might have been worth the added intrigue.

The Appalachian Trail is more than a trail. It's not simply a path through the woods. To describe it simply as a "trail" is an injustice. And to call it a hike? That's also pathetic. It's not just a hike. It's an ADVENTURE. It's a roller coaster ride, a sweaty, cow-pie avoiding amble through fields, a scramble over boulders, an amble along riversides, a ascent up and over mountains, a slide over muddy and slippery roots.

Of course the trail is dangerous. LIFE is dangerous. But a life you don't live is a waste. If you're lucky enough to get a long vacation and the prospect of a long hike doesn't make you miserable, you should consider it. Most likely, it could be the most amazing journey of your life, an exploit worth months of wet socks and sausage toes, filthy hobos, instant mashed potatoes, (and did I mention bears?).

Like I said, I used to spend a lot of time trapped in worlds of fiction. Eventually, I was lucky enough to learn that even though reading someone else's story can be a tremendous adventure, living your own is even better. You can read as many books as the library holds, but you only have one life. Who wants to die without having their own story to tell?

If you are fortunate enough to have the time, funds, and physical capability to make your own adventure... well... how could you live without it?

~ ~ ~

Danielle grew up in Connecticut, moved to Providence, Rhode Island for college, missed the wilderness and decided to major in Environmental Studies. After an internship in Acadia National Park she moved to Seoul, Korea to teach English. Returning to Acadia she learned of a friend's thru hike and decided to commit to a hike of her own in 2015. It was the best decision she ever made!

Chapter 6 Five Million Steps Times Two
Cynthia and Woody Harrell aka N & X Trovert

An old trail adage says if you spend forty years preparing to thru hike the Appalachian Trail, there's a good chance your training will peak too soon. However, we found a four-decade delay doesn't mean hikers holding an AARP card can't enjoy a wonderful, life altering experience, equal to, or even better than, that of the many "Young Machos" with whom they will be sharing the AT.

During the summer of 1975, I was making a living giving first person history programs portraying airplane inventor Orville Wright in his Kitty Hawk work shed. Early that season, I met my future bride when she approached the information desk at the Wright Brothers Memorial, and I inquired, "May I help you?" (At the time I failed to realize two score years later I would still be dealing with the ramifications of that simple question.) Over the course of the summer, Cynthia and I began talking about a thru hike of the Appalachian Trail. Initially, nothing came of it, but the idea always at least simmered on our back burner, even as kids, careers, and life in general got in the way.

During the 1990s, a cross country car trip listening to Bill Bryson reading "A Walk in the Woods" brought the Appalachian Trail back on our radar. I began telling everyone, "When I retire, I am going to hike the AT." After repeating that promise enough, it's easy to become committed to (trapped by?) the idea. However, I also found the reality of that statement became a tremendous incentive to keep on working! Finally, as I approached age 65, Cynthia and I both realized it was now or never. We decided it was time to shift careers and become Professional Walkers: We would get up. We would walk. That is what we would do. Then the next day, we'd get up and walk again...

From our earliest planning, it always seemed to us if you're going to be hiking in pairs, you'd need trail names that somehow went together. So, in preparation for our thru hike, acquiring matching trail names became a high priority. This put us into direct conflict with an Unwritten Rule of the Trail: trail names must be bestowed or conferred by fellow hikers on the basis of an observed character trait or some trail incident. But depending on other people to come up with two superior connected names seemed leaving a lot to chance.

Thus choosing our own trail names became a critical part of our AT preparation. And for a long time, it proved a futile task. In spite of our best efforts, nothing seemed an ideal fit. We arrived at the Hiker Hostel just outside Dahlonega for our final pre-thru hike, hundred-mile shake down sans perfect trail names. We agreed, well, okay, let's pull several names from our short list of possibilities, and we'll try a different pair out each day to see if we can find names that actually work on the trail. And then somehow, the next morning when we woke up, low and behold, we discovered miraculously overnight, we had become the Troverts – "X" and "N." How it happened, we can't explain. It's a mystery. But then it would be hard for us to deny to anyone who knows us, that those trail names aren't a great fit.

When you say "2,000-mile hike" quickly, the phrase rolls easily off the tongue. As does talk of a "six-month hike". But when you start to consider exactly what is involved, the magnitude of the AT becomes pretty staggering.

Back in my running days, I had occasionally trained to knock out 25 miles in two and a half hours. However, the AT isn't the typical paved, mostly smooth, one-day road race. And on top of the miles of roots and rocks, we'd face enough elevation gain to climb from sea level to the top of Mt. Everest – 17 times. Although the 20-somethings we'd see on the trail could easily go faster, at our stage of life we'd have to be content with about two miles per hour. With that pace we were still looking at walking a half marathon every single day, and needing to keep that effort up for five and a half months straight, over 166 half marathons in all!

Winter weather in the Great Smoky Mountains often makes an early start on the AT a recipe for disaster. January and February 2012 were both mild enough we might have gotten away with such an early departure and not frozen to death, but work obligations kept us off the trail until April was one third gone. Thru hikers begin to trickle onto the trail in January. In February it becomes a veritable torrent and finally into a flood of people flowing north from Springer Mountain. For those of us still counting our start date in months to go, we faced a depressing time, as we started to feel vaguely left behind. After the arrival of spring, we were squandering precious hiking time while the days are longer than the nights. This dilly-dallying would have to be paid for if our hike extended past 9:49 EST on September 22th and the Autumnal Equinox.

When April 11th finally arrived, our late start meant we would always be keeping one eye on the fall calendar and the October 1st closing date for Baxter State Park at the AT's northern terminus. Given our proposed schedule plus a little time off to attend a July family wedding, we had left ourselves no margin of error.

In hindsight, we would not have done it any differently. We avoided both the worst of the weather and the early AT crowds. Having prepared with no longer than daily 3.5-mile pre-dawn, pre-work walks around the neighborhood (at a fairly brisk 14 minutes per mile pace, but still only three and a half miles), we did not have our trail legs when we began. But our sense of urgency did focus us on the necessity to go to bed every evening 13 miles further north from where we had woken up that morning. A steady pace wins the race.

While living on the Grand Canyon's South Rim as young parents, we made frequent overnight hikes in the Canyon and nearby national parks. I have vivid memories of carrying prodigious weights (mostly bottles of water) with two small boys in tow on treks into Arizona's arid backcountry. As we began our AT planning, I knew the days of hauling such heavy packs were a long way in our rear view mirror.

More recently, all our camping had occurred as we checked off visits to the 22 national parks in Alaska. Traveling "America's Last Frontier" by kayak, raft, and bush plane, the total weight of our camping gear had not been a concern. However, when we went out for our first multi-day practice hike on the AT (an early season 50 miles north of I-40), we found the hand full of thru hikers already on the trail were all zooming past us. Pretty quickly we decided we had to upgrade to some lighter gear if we had any hopes of succeeding on a long distance hike.

Adhering to *Trail Wisdom*, "Pack weight is determined by fear." So, we sought to overcome any irrational concerns of being cold, of being hungry, of being out of touch, so we could carry the fewest clothes, the least food, and the fewest electronics possible. We retired two perfectly serviceable backpacks to acquire newer models, each three pounds lighter. We likewise replaced our "easy set up in the rain" Alaska model for a newer, lighter, three-season tent.

The Christmas preceding our start, we gifted each other with a kitchen scale which measured down to 1/20 of an ounce, and began to attack our preliminary gear list in earnest. We chipped away at pack weight, cutting it down to size one fraction of a milligram at a time. Sharing communal gear like the tent, stove, and water filter (a big advantage for a hiking couple), we eliminated everything we possibly could, and got our base weights down to 20 pounds apiece without food. We thought we were ready to go.

Then we found if you finalize your gear list too early before your hike, sometimes the pressure of "gear creep" will insidiously start to add small items of inconsequential weight back to the list, ounce by ounce. And that's not even considering last minute advice we received, such as, "There is no bad weather, only inadequate clothing." Maybe we need to add one more pair of socks?

Although we had decided to skip the Approach Trail, as a hike prologue we stopped at the Visitor Center at Amicalola Falls State Park to register as AT hikers #'s 839 & 840 for the year. So, yes, we were officially making a late start (although we had no idea what per cent come to the VC to sign in versus those who proceed directly to the Springer Mountain parking lot). We were Numbers 13 & 14 to sign in for the day (with pack weights of 22.4 and 23.6 pounds). We noted most folks on the register did not yet have trail names.

We had set aside enough funds so our AT budget allowed us a strategy we referred to as "Roughin' It Deluxe", which we decided to incorporate right off the bat. Our first three nights were all spent at the Hiker Hostel, giving us the luxury of two days of "slack packing" before we began carrying our complete gear load at Neel Gap.

Twice we left for the trail showered, with full stomachs and a good night's sleep and with a reservation so we could come back and do it all over again. Yes, no need to carry a tent or a stove or a change of clothes or almost anything: we would be slack packing! Light of heart and light of backpack, we were shuttled to and from the trail.

Even after these 30 miles to "ease on into" our hike, the weight of our full backpacks on Day 3 caught us by surprise, and immediately earned our packs the trail names of "Shock" and "Awe".

During our earlier section hikes, we began rating our mornings with a "one" or a "two" crow bar designation, depending on the number required to pry ourselves out of our sleeping bags. Day #4 was already up to three crow bars (technically three goose-neck wrenching bars). Only copious amounts of Vitamin I (Ibuprofen) enabled us to haul our sore butts out of bed. But after a good breakfast

85

and a little moving around, we were ready to face the challenges of the day.

However, even with two people to tackle camp chores, we found ourselves very inefficient. And foregoing a hot breakfast didn't seem to gain us the extra time we'd anticipated. (Although, it did allow us to complete the entire trail without ever once tasting instant oatmeal!) Plus, the bulk of our gear magically expanded each night, making increasingly difficult the task of stuffing it back into our packs the following morning.

The slow re-packing put us in a daily hole from which it was tough to recover. A truth of the trail is there are some miles that are just not like the others. A majority of AT shelters are nestled down snuggly in the gaps of ridges and mountains to be close to water sources. What that means is, whether you're a north bounder or a south bounder, when you leave camp in the morning, the first thing you're going to do is hike uphill!

But within a few days, we happily found ourselves getting into a rhythm. We were very comfortable walking together, and the climbs started coming a bit easier. Remembering the "narrow for chosen company" part of Harold Allen's trail description, we were "beckoned upward in body, mind and soul". Our appetites were off a tad, but we had heard that was expected when just starting out. The weather had been absolutely phenomenal and we felt very lucky for that. And on Day #6, we said "Hello" to North Carolina! (Our fifth state, if you also counted the state of excitement, the state of euphoria, and the state of exhaustion.)

The climbing got serious as we approached the 5,250 foot Albert Mountain (240 feet lower than the previous day's Standing Indian Mountain, but a much tougher climb). We found the approach from the south to be extremely steep and rocky, but a fitting location to celebrate completing our first one hundred miles. Albert (plus its mandatory fire tower ascent) certainly put into perspective our training on Mississippi's Woodall Mountain (the state's highest natural point, reaching an imposing summit of 806 feet above sea level).

Nearing the highway to Franklin, North Carolina, we were looking forward to our first zero day, and all of the amenities of a trail town. If only the weather would cooperate, we would be right on schedule for a comfortable hitch to town. However, around 4:00 p.m. light rain arrived and out came our pack covers and gaiters. As we began the climb out of Wallace Gap, weather-wise, all heck broke loose: Claps of thunder directly overhead, lightning within several hundred yards, and finally a pelting of pea-sized hail began to hammer the Troverts.

We were cold and wet when we finally heard the traffic ahead on U.S. 64. With darkness now approaching, we didn't relish the prospect of a long wait trying to hitch a ride in the storm. Thankful for a signal, we frantically got on the cell phone to try and arrange a shuttle. The best we could do involved an hour's wait, as our potential driver was in the middle of cooking her dinner.

Now here's where the story gets truly amazing. Just before we hiked out of the woods overlooking the Winding Stair Gap parking lot, the Sutton family pulled up in their big pickup truck. We could see them looking up to where the AT emerged from the woods. They later explained, "We always look for hikers any time we come over the gap." They apologized for having space only in their exposed truck bed, but asked if they could give us a ride to town. With the rain now really starting to pour in buckets, we tried to intercept our driver Kathy, as the Suttons handed us juice and snacks. N Trovert asked for a recommendation on a local delivery pizza, as he thought if we ever made it to a dry room and hot shower, we'd never want to leave for the rest of the night. Driver Kathy got back to us with a text to say "God provides" and she would go back to fixing her dinner.

We hunkered down like two wet pups in the back of the open pickup for the exciting ten mile ride into Franklin. When we pulled into the Budget Inn, a lady came rushing out to say there'd been a mix-up and in spite of our reservation, she had no rooms left. However, there were still rooms at the Sapphire Inn just down the street. Before we could get too upset, the Suttons said they'd be glad to take us the extra mile to our new home for the night.

As they dropped us off, Mr. Sutton said unless we really had our hearts set on pizza, their brother-in-law ran a great place that had a homemade turkey and dressing special, and they'd be happy to go pick up some dinner for us. They took our order for vegetables, and were back in a matter of minutes with homemade pie thrown in for dessert! For all of which they'd accept no money. From the verge of hypothermia to the lap of luxury with some of the best "comfort food" we've ever tasted. Talk about trail magic and trail angels! The rain and lightning continued all that night, and we were very glad to be off the trail, even though we worried about our fellow hikers still south of us who decided to tough it out with Mother Nature.

Which led to our first Zero Day! Are there any two prettier words in the English language? We won't say we didn't do any walking, because we did, (even some uphills) but instead of counting on our AT total, these miles were all spent in doing town chores around Franklin. Well, OK, maybe there were a few "want to's" thrown in there with the "have to's" like X's McDonald's coffee, right across the street from our motel. After a diet of pop tarts, a Sausage McMuffin with Egg tasted truly decadent!

Zero days in town were unfortunately followed by uphill climbs with heavier packs. We tried to be positive, even with the foggy weather, which after all was an opportunity for us to "Wash our faces in the clouds". Within a month, we were getting our trail legs, and definitely losing weight.

With practice, we were now able to cut our morning prep time by a good half hour, and we tried to be on the trail as soon after sunrise as possible. We didn't always make it, but our standard goal was "8x10" (eight miles before a stop for a ten o'clock treat). The only downside to these early starts was one of us often spent the first hour or so using our face to clear out spider webs from the trail before a young macho marching by at twice our pace took over that thankless task.

We would later see zero visibility, 80 mph wind gusts, and hard rain with temps just above freezing on the slopes of Mt. Washington, but our closest call with hypothermia came on Clingmans Dome, the AT's highest point. The day before had seen an end to 36 hours of the "sunlight deficiency blahs" with the return of shorts and t-shirt weather. For X Trovert, the sun shine meant it was off with her toboggan and on with her more characteristic pink bandana headband.

In a rush to cover the nearly eleven miles we needed to meet our ride at Newfound Gap (for another night off the trail and more laundry & food resupply), we got an early start from Double Springs Gap Shelter. Hiking in on-and-off-again showers, I was hesitant to spend time donning and removing raingear. Approaching the summit conditions worsened with a sharp increase in wind velocity and pelting hail.

Trying to pull on gloves and dry clothing, I found my cold hands could not manage my buttons or zippers. We retreated off the trail to a lower visitor parking lot looking for a warm building, but we found nothing open. There was no place to hide from the wind and rain. X Trovert used our tent fly as an emergency wrap around windbreak so I could trade out clothing, then we retraced our steps back to the AT. We had squandered an hour of hiking time, but had avoided serious hypothermia issues. It was one morning I was glad for an extra set of hands. By evening, we were not the only hikers the continuing bad weather had sent scurrying down the hill into Gatlinburg.

The Appalachian Trail is one of 400+ units that make up our national park system. The National Park Service is traditionally one of the public's favorite federal agencies. Yet, if you were to survey AT hikers, most would list their time in the Great Smoky Mountains and Shenandoah National Parks as some of the most regulated and least enjoyable stretches of the trail, rivaling their later negative experiences in New Hampshire with the hut system of the AMC (Appalachian Money Club). The rules are well-intentioned attempts to deal with the AT's popularity and resulting overcrowding. But that makes the regulations no less irritating to folks who find the freedom of the trail to be one of its greatest appeals.

As a break in the weather flushed the hiker backlog out of Gatlinburg and back on the trail, we heard many hikers commenting on their desire to get beyond the Smokies. They especially wanted to leave behind the park regulation requiring AT hikers to stay in the Park's overcrowded shelters. After hearing tales of 52 hikers just behind us spending the night at Icewater Shelter (a structure designed to house 14), it was hard for us to disagree.

After crossing north of I-40, overcrowding became less and less a problem. Hiker attrition certainly played a role in this. At Harpers Ferry we signed in as NoBo hikers # 614 & 615 (down from among the mid 800s at the start), and at the base of Mt. Katahdin we registered as # 538 & 539. With each of the five million steps along the way, we came to appreciate luck has to be a key factor in a successful AT hike. Fortunately, we relied on a continually growing network of folks who helped us reach our goal. We started with friends and coworkers, especially from the park service and nature center communities. We added relatives of friends, and then friends of friends. "So and so lives fifty miles further up the trail. I'm sure he'd love some company. Let me give him a call." And three days later here would be someone waiting as we came out from the woods to act as our host for the night.

Sometimes we imposed mightily on these help providers. From the start the Troverts' AT hike motto had been, "The dogs bark, but the caravan moves on." But while hiking through the quiet corner of Connecticut, a medical hiccup came at me out of left field.

We had been staying just off the trail with a friend's parents who made their entire two day priority fattening us up and supporting some slack packing along the Housatonic River. However, as we were scheduled to leave that Friday morning, after a few days of increasing fatigue and a growing tenderness near my left shoulder blade, I woke up to find an old issue with a skin abscess had returned with a vengeance. Yep, looking just like the nasty pictures X Trovert googled on her iPhone, there was a large boil on my back.

Our hosts pulled some strings and got their primary care physician to work me into his schedule. (The doctor was much more interested in our point of origin than the uniqueness of my condition. "In all my years of practice, you're only my second patient from Mississippi." X Trovert also pointed out that I had walked 1,477 miles to be able to see him!) The doctor's proposed treatment -- lance the boil, use frequent hot compresses to aid the drainage, and wait 48 hours for the results of a bacteria culture. It all sounded simple enough, unless you're on the AT, where water, much less HOT water is not always available. Like it or not, we were going to be in town for the next few days.

Our hosts informed us our revised motto had now become, "Sometimes when *Staphylococcus aureus* bacteria comes to a head, the whole caravan has got to take a few days off." They had invited us for a night, but before the doctor's final "Okay", we wound up having to stay for a week. Their gracious hospitality went way beyond what one might hope for from a best friend, not a stranger, and may have taxed even the original good Samaritan.

Since the turn of the Century, the Internet and especially the smart phone have wrought profound changes for those walking the AT. A thru hiker now has instant access to the latest weather radar, or contact with nearby stores, hostels, and shuttles. A proliferation of blogs is another phenomena of the digital age, enabling any would-be hiker to master ahead of time all the nuances of trail culture, and be as prepared as possible. We certainly learned a lot from reading several 2009 online journals such as Wags, Gangsta, and especially Tagless and Tag-Along, a hiking couple with whom we could easily relate and identify.

To familiarize ourselves with the challenges of maintaining an electronic AT journal of our own, we

began posting our own entries on January 1st. Our first
hint of notoriety came when we were contacted by a
family near the Maryland/ Pennsylvania line who wanted
to use our hike as a unit for homeschooling their children.
(We later met these folks to enjoy their trail magic at Pen
Mar Park.) And we never ceased to be tickled when
greeted as "the Troverts" by passing strangers who
recognized us from reading our story on
TrailJournals.com.

We awoke on the 4th of July to the quandary of either
attempting one killer day into Port Clinton (especially
when factoring in reports we'd gotten about the infamous
Pennsylvania rocks in the upcoming section) or two easy
days into Port Clinton, if we split up the distance around
the next shelter and water. Finally, even after a delayed
start from a threat of rain, we decided, "OK, it may take
us until after dark, but somehow we'll trudge on into Port
Clinton."

Not two miles down the trail, we met a man walking
toward us. Looking much too fresh to be a AT hiker, he
called to us by name. It seems Dave Martin enjoys
following folks on line, then coming out to offer
assistance when they pass through his area. He quickly
reversed course and walked back with us to his parked car
at PA 183.

At the road, he offered us a cold soda.

Dave: "Not yet eight o'clock, probably too early for a
Coke?"

N Trovert: "Never too early for a cold soda." (as I quickly
guzzled the entire can)

Then he blew us away by offering to slack pack us the
remaining 14 miles to Port Clinton! Are you serious???

We jumped all over his kind offer, and filled two large
trash bags in his trunk with everything we didn't need for
the next seven hours. We headed back into the woods
with both a figurative and literal spring in our step!

A few hundred yards into the woods, the adrenaline rush
wore off, and we stopped to look at each other. "Did we
just hand over all our worldly possessions to a perfect
stranger? What were we thinking?!" At that point, all we
could do was continue on.

It's amazing what a difference a little weight can make.
With our hiking time shortened, we didn't need to stop to
take on water, which saved us even more time. And
keeping our balance on those sections of nasty
Pennsylvania rocks was not as painstakingly slow as it
would have been with full packs. Why it was like we were
living the lives of two young macho hikers! All of a
sudden we were loving the AT in Pennsylvania. Bill
Bryson, you don't know what you missed!

While we had a much smoother time than we had
expected when we first woke up, the AT from Eagles
Nest Trail to Port Clinton is not easy. The trail downhill at
the end is amazingly steep and not something we'd want
to face on a rainy day. But just past the train tracks at the
bottom, as promised, there was Dave Martin with our gear
and more cold drinks! And after we inhaled those Cokes,
he would not quit until he had taken us safely to our
motel.

Barely past the AT's half way point, we had accumulated
a hundred stories of amazing trail magic, each and every
one appreciated, but we were sure our encounter with Mr.
Martin could not be topped. And it wasn't -- until we got
to Bennington, Vermont.

Steve Labombard had to be our most consistent and conscientious journal reader. We didn't have relatives, much less total strangers, who followed our ramblings as closely. Steve had been on the lookout for us with trail magic in the Bennington Gap, but we somehow missed him as we passed through. Later, we heard we had missed him by about a half an hour at Mad Tom Notch. (Our pace was fast enough that we were through earlier than he had expected us.) Have to admit we got the collective "big head" when we got an email from Steve confirming he had been on the lookout for us!

A few days later, we were running low on energy half way through a long day into the town of West Hartford. But as we came down to the obscure gravel Joe Ranger Road, that all changed. First we saw a vehicle with an open hatch, then some coolers, then a small folding table, and then when the man said, "I know you're the Troverts!" We knew it was Steve Labombard and his canine companion Tank.

Unbeknownst to us, earlier that morning, in our rush to start our next climb, we had missed Steve by less than five minutes at Woodstock Stage Road. However, he met the rest of our current bubble of hikers, still at the road, and got an update on our plans. After taking care of everyone else, Steve had packed up shop and moved to intercept us at Joe Ranger Rd. He plied us with sodas, homemade cake, and fresh fruit, while firing up his camp stove to make us Sloppy Joes. We were overwhelmed! On the fourth try, his timing was perfect, as we were much more in need of a break and a re-fueling then we had been at the previous near misses. It was great to finally meet Steve and we couldn't thank him enough for the bounty he bestowed. Only on the Appalachian Trail could this sort of thing happen.

On the AT, the really big news, like the tragic drowning death of the early season hiker, Parkside, spreads quickly up and down the trail. But for most of the 2,000 mile hike, your little world (the group you keep up with) is the bubble of hikers traveling just ahead and just behind you. For example, LongStride and SilverGirl, another couple hiking in 2012, always remained ahead of us, and we met them only via the internet. On the other hand, Manula and Tree Trunk, a couple from upstate New York, started just ahead of us. We had frequently heard about them as we hiked through Virginia, long before we first met at a stop while looking down at the Potomac River. We took an immediate liking to them, and as we were hiking at about the same pace, we thought we'd see them a lot more. However, our encounters, while always delightful, proved to be sporadic. We always seemed to be just ahead or just behind. Then as luck would have it, we would camp beside them in Baxter State Park and finish our hike together the next day, celebrating on the summit of Katahdin.

While in Connecticut, we started bumping into Kleen X from South Dakota, who was hiking solo since her sister had left the trail early on. We climbed Bear Mountain and beyond in her company, before we lost contact when we hiked past a shelter to rough it deluxe at the lodge on Mt. Greylock. Then ten days later, while both Troverts were having a very lethargic morning, who should walk up but Kleen X, back after a few days off the AT. This gave X Trovert somebody new to kibitz with, and the result was a dramatic increase in our pace. Coincidentally, the Sun came out at the same time. Across Vermont, we were frequently stopping at the same campsites as Kleen X, and were often joined by Castaway, another member of our bubble. By the time we reached Mt. Moosilauke, we had become an informal gang of four, and shortly after, decided to pool our resources as a formal team, sharing communal gear to reduce weight. Anticipating colder weather in the final 500 miles, we had had our Base Camp Manager return the heavier clothes and sleeping bags, we had traded out for lighter summer outfits back in Virginia, so this arrangement worked out very well. Lasting new friendships from the trail have been a big bonus of our AT hike.

When you're burning calories as quickly as we did on the trail, an obsession with food is understandable. A listing of our most memorable meals by itself would form a thorough summary of our hike: home cooked meals on some of the 32 nights we spent in someone's home, the "Deli tour of New York" (where after crossing the

Hudson River, for three straight days we managed lunch stops with amazing sandwiches, all within a half mile of the trail), AYCEs ("All you can Eat") wherever we could find one, "work for stay" leftovers in the Whites. Hot Springs, North Carolina, had juxtaposed two very different dining experiences. We had anticipated a great gourmet vegetarian meal and stimulating conversation at Elmer Hall's Sunnybank Inn, and the evening did not disappoint (special kudos for the pecan pie).

The next morning found us across the street. Our great breakfast at the Smoky Mountain Diner brought back memories of two still newly-wed thru hikers from Chapel Hill we met there years earlier. They certainly had their act together, and over the next several months we had enjoyed following their progress on their attractive AT website. We would have bet the next month's mortgage payment on their ultimate success. But then in Pennsylvania, they discovered, for several weeks, each had lost their zest for the hike and was only continuing so as not to let the other down. They commented that it seemed like all the thru hikers were wearing blinders, as they trudged on, fixated on Katahdin, while every section hiker they encountered seemed to be having the time of their life. With that realization, they stopped and headed back to North Carolina, confident they'd return to finish when the time was right.

We had wondered then whether we would be able to keep our hike a "want to", rather than a "have to". Barely begun, we were off to a great start, but could we keep it going? Over the rest of our hike, we found a big key was concentrating on the journey as much as the destination. We didn't start worrying about the next section until we'd finished the current one. "Never quit on a bad day" was great advice. With two people hiking together, the chances of both having a bad day together was less likely. A sign we saw at Mountain Harbor Bed & (World's

Greatest Ever) Breakfast said, "If you're lucky enough to live in the mountains, then you're lucky enough." Following the AT's long green tunnel for 168 days, we felt we were awful lucky.

~ ~ ~

Following rewarding careers in historic preservation and environmental education, Woody and Cynthia Harrell used their AT thru hike as an ideal transition into retirement.

The co-recipients of Time Magazine's 1966 "Person of the Year" award first contemplated the trail while living on North Carolina's Outer Banks, well before the AT's recent exponential rise in popularity. Later, while working for the National Park Service on the rim of the Grand Canyon, the Harrells enjoyed many chances for literally "take your breath away" short backpacking treks. However, their opportunity for genuine long distance hiking arrived only after decades of repeated postponements, due to life's little distractions, such as jobs, kids, and hobbies.

Post-trail, the Harrells have lowered their daily mileage total, while increasing their speed, employing running as a way to maintain their lean AT ending weights. Both have raced in nationwide long distance events, topped by the prestigious 26.2 mile Boston Marathon. After five and a half months spent walking America's longest national park end to end, the Harrells have returned to Woody's long standing quest to visit all 400+ national park units, a task which never seems to stay completed, as more and more parks continue to be added...

Chapter 7 If Turning Left is Wrong I Don't Want to Be
Right   **Jackie Kuhn aka Left Turn**

How does a 53-year-old woman find herself on a solo thru
hike of the Appalachian Trail? My motivation was pain
and grief, a plain and simple mid-life crisis. Around
Christmastime 2013, I was floundering in a new job as a
staffing manager at a homecare agency. The first time I
had to let an employee go, I queried Google: "How do I
fire someone?" I was in over my head. Brushing my hair
one morning, I discovered a sizable bald spot above my

right ear. I had alopecia auriata, an inflammatory condition commonly caused by stress. A few days later, a friend went a little nuts and quit her job and when I refused to lend her money, she tried to blackmail me.

I'm stressed out, balding, and more than a little afraid of this crazy woman, when I get a break up text from the guy I was dating that said, "I'm out. Keep the mower." (He had lent me a broken lawn mower that required carburetor spray and finessing of the choke to get it started. Not a great parting gift.) This trifecta of woes tested my belief that people are good and the world is a kind, nurturing place. I felt sad, lost, confused. I took a break from crying in my room to attend a talk given by a friend. During his presentation, he said, "When you hit a brick wall in your life, it's time to make a left turn." Boom! I went home from that event and paced back and forth in my room, thinking, "I need a left turn. What is it going to be?" I glanced at a book on my nightstand, *Hiking Thru: One Man's Journey to Peace and Freedom on the Appalachian Trail*.

I was having difficulty adjusting to flat Florida after relocating there from the west coast in the fall of 2012 on the heels of my divorce. I missed California's mountains and I envied the author Paul Stutzman, who walked away from his grief to hike the AT from Georgia to Maine. Hey, wait a minute! I could do that! I had romantic notions about hiking the AT one day. Could this be the left turn I'd been looking for? I ran the idea by my daughter Vanessa.

"I'm thinking about hiking the Appalachian Trail, from Georgia to Maine, it will take me six months. Do you think I'm crazy?" Without hesitation, she replied, "You'd be crazy if you didn't, Mom." Man, I love that girl.

So, I spent two months planning and preparing for the trip, and I made those plans official in late January 2014 when I announced my intentions on Facebook. I felt I had to be accountable and would follow through, because, you know, I said it on Facebook. Scared to death, I began my walk on March 9 at Springer Mountain, Georgia, the southern terminus of the AT. I chose my own trail name: Left Turn.

The walk from Georgia to Maine challenged every fiber of my being. I trudged through much of my hike, yet on other days, I floated above this difficult world, wild and free, living in harmony with nature. But, then there were those trudging days. Looking back, surviving the hardest days are what made me feel fully alive, part of nature and the animal kingdom, and united with my fellow travelers in a quest for the intangible. The hardships are what brought me the most unexpected and thoroughly delightful surprise of my hiking experience, the thoughtful kindness of strangers who appeared as my trail angels. These are people who live near the trail and provided rides to town, slack packed, and even invited me to stay in their homes. There were many readers of my blog on *trailjournals.com* who followed my hike day-by-day and sent me encouraging messages. Those guest book entries kept me company on my loneliest days. The generous outpouring of support and heartfelt encouragement helped keep me determined and focused.

Many people ask me what scared me the most. Did I carry a gun? Was I afraid of people or animals? No, what scared me the most was weather, with insect bites coming in a close second. Five weeks into my journey -- a memorable day because it was Tax Day, April 15 -- I hiked across the Roan Highlands by myself in a ferocious blizzard. I grew up in Chicago, the Windy City, but never encountered freezing sideways rain, sleet, hail, snow, and winds like that. Winds strong enough to knock me over,

and those highlands go on and on, with one false summit
after another -- Round Bald, Jane Bald, Grassy Ridge
Bald, Big Yellow Mountain, Little Hump Mountain, and
Hump Mountain with no trees and barely a rock to hide
behind for shelter.

It took 100% of my energy to stay on my feet and put one
foot in front of the other. I had to lean into the wind and
then when it let up for a second I would almost fall over.
I'm sure I looked drunk. And, I probably sounded drunk
as I yelled at the heavens to please give me a break! The
vegetation was frozen and had a layer of ice covering the
side the wind was hitting, as did I. My eyebrows froze
and everything hanging off my pack had a crunchy
coating on it. I reached the safety of Highway 19E and
slept in a warm bed at the Roan Mountain Bed &
Breakfast.

I took the next day off, and when I started hiking north
again, I made a conscious decision to hike my own hike at
an enjoyable pace. Up until that point, I was worried that I
was hiking too slowly and couldn't keep up with the
mostly 20-something hikers who were bragging about
their 25 mile days. That evening I wrote in my journal:

> Today was a very spiritual day for me on the trail.
> I have experienced quite a bit of pain and
> discomfort on this journey, physical and
> emotional, but it has not been for naught.
> Choosing a camping spot, making all of these day-
> to-day survival decisions is no small thing for me.
> They are accompanied by great self-doubt, but
> with every decision and it's eventual outcome that,
> yes, I have survived another day, comes a bit more
> self-assurance, a bit less longing that the past had
> worked out differently so I wouldn't have to be out
> here right now testing myself at this great level.
> I'm coming around to be genuinely glad about the

103

circumstances that have brought me to this very moment in my tent in the woods typing this journal entry on my iPhone with cold fingers. AT thru hikers are an odd lot, and I'm right there in that bunch. One of my fears about this trip is that I would hike the trail but stay the same...come home unchanged. I am beginning to see that is not going to happen. I'm growing more than muscles on this hike. More will be revealed.

The next day, I received a message in my trail journal from Melanie from Pennsylvania:

> Jackie, I truly love reading your journal. It is down-to-earth truth. Your soulful truth... unvarnished and unglorified. You have learned early on, what so many thru hikers learn much later (myself included), or never learn -- to experience the moment and SEE (inwardly). I am so happy for the hike you have chosen for yourself Jackie. You have come so far, and will continue many miles, with many beautiful experiences awaiting you. Melanie of "The Bobsie Bums" AT '88-90, PCT '91-'92

I received over 800 messages from kind strangers like Melanie and, oh my, I can't begin to express how important they became to me, a balm that eased my woes and pains at the end of each day. I hiked with my phone in airplane mode to take pictures, but looked forward to the time each day when I would turn it on and search for a signal so I could upload another day's journal and photos, as well as download those daily words of kind encouragement.

During the planning phase of my hike, setting up an online trail journal was important to me for several reasons: to keep family and friends up-to-date about my whereabouts and experiences, to remind myself where I'd been and what I'd seen and done (because I have a really bad memory), and to share logistical details with other would-be hikers. Updating my journal daily was my way of giving back to the hiking community. Reading previous hikers' accounts helped me prepare and also helped me believe I could actually accomplish an AT thru hike.

Wow, I got so much more out of it than I ever imagined! Not only did readers write wonderful supportive comments in my guest book, but some of them actually met me on the trail and invited me to eat at their tables, shower in their bathrooms and sleep in their guest rooms. Remember, I started my hike to get over my disappointment in others and myself. The AT is a magical place that gives those who walk it exactly what they need at that moment in time. What I needed was to restore my faith in humanity and my faith in myself. The trail delivered. In mind-blowing proportion. I eventually felt grateful for the hard knocks that gave me the desperation to attempt a crazy thing like a six-month mountain walk.

I benefited from much anonymous and spontaneous trail magic in the form of coolers of sodas left at a road crossing, a pop-up barbeque at Tray Gap in Georgia hosted by two guys named Bob and Jeff who told me they fed hikers because, "We love the Lord, and we love to serve." I received countless rides into town from strangers who went out of their way to help me. I would love to tell you about each and every one of those experiences, because they were all amazing in their own way. But, for brevity's sake, I will just tell you about a few instances of kindness that I received from folks who were reading my trail journal.

Three days into my journey, I was descending Blood Mountain with only 2.4 miles to hike to Neel Gap and the

Mountain Crossings outfitters and hostel, when I encountered a couple day hiking. The husband was taking a picture of his wife, so I offered to take their picture together. I almost had to twist the guy's arm, but they posed for the picture and then I continued down a steep granite path that I thought was the trail. Little did I know that they had been standing in front of the double blaze indicating that the trail curved to the left into the woods, so I missed the turn. I scrambled down a hard-going Rocky Mountain face and realized that I hadn't seen a white blaze in awhile. I had heard the hike down Blood Mountain was challenging, but I had a feeling it wasn't supposed to be this hard. I retraced my steps until I found a white blaze, scrambling and crawling up the rock face in the early afternoon sun. I drank almost all my water and sat on a rock to recover once I reached that perfect photo point where I lost the trail.

The hike down Blood Mountain was indeed very challenging, especially after tiring out my legs and feet doing all that unnecessary rock scrambling. I was still feeling pretty foolish and regretful of my difficult detour when I bumped into a mother-daughter hiking duo, Flash and Noodle, who recognized me from my Trail Journal and offered to share their cabin with me at Blood Mountain Cabins -- such good fortune, since the hostel at Neel Gap was booked solid and a cold rainy night was fast approaching.

While relaxing in the cabin, the legendary Miss Janet stopped by to take a shower and eat; she offered me a slice of cheese and salami from the meager portable pantry she kept in her van. She beamed with an inner light. On day three my body was aching and I was questioning my gear, my choices, my planning; and was sorely lacking confidence about my ability to thru hike. Miss Janet smiled and reassured me that she had no doubt I would finish my hike – she conveyed mercy and kindness, the very qualities I needed to give myself. I felt like a winner when I walked out into the freezing rain the following morning to take on whatever the trail presented. Good thing, too, because that was a brutally cold day.

Weather continued to be an issue for me. I started my trek about three weeks earlier than most thru hikers, to ensure that I could finish before snow fell on Mt. Katahdin in the fall. After hiking in some cold rainy conditions, I was actually looking forward to the snow that was forecast for the day I entered the Great Smoky Mountain National Park. Oh, was I wrong to think snow would be better than rain! I developed silver dollar sized blisters on the back of my heels because my boots were slipping off as I was forced to cross the Smokies by what is called "post holing". Each step in the soft snow makes a narrow, deep hole like one would make to sink a fencepost. The only way to make forward progress is pulling each half-buried leg straight up out of the snow before taking the next step. This took all my energy and my loose boots and wet socks rubbed my heels something fierce. A woman named Connie (trail name Dragonfly), who lives in Knoxville, Tennessee was reading my trail journal. She offered to pick me up at Newfound Gap and let me stay at her house for a night or two. I wasn't quite sure what to think of staying at a complete stranger's house, but I was pretty miserable so I took a risk and agreed to meet her.

We had a plan to meet at Newfound Gap at approximately 7:00 .p.m, but I underestimated the time it would take me to summit Clingman's Dome, the highest point on the AT, and hike 18 miles in the snow to get to Newfound Gap. I didn't reach our meeting spot until 10:00 p.m., hiking with a headlamp on the darkest night of the new Moon. Connie waited for me until 9:30. Since neither of us had a cell signal that afternoon, she figured I was unable to make it down and that I probably stopped at the Carter Gap Shelter for the night. I tried sending a text, "Please don't leave, I'll be there soon." But I had no signal for the eight miles from Clingmans to Newfound.

Connie and I were finally able to exchange texts when I reached the desolate, dark, windy, cold parking lot. Connie texted that she was almost back to Knoxville but that she would turn around and come back to get me. At this point, I had been hiking for over 12 hours in grueling conditions and was physically and emotionally spent. I now know that there is a good-sized clean bathroom near the parking lot where I could have found shelter for the night, but because of the pitch black night of the new Moon, I had no idea I was standing 50 yards from an unlocked building.

A Park Ranger named Ellen pulled up while I was shivering and crying in the Newfound parking lot and asked if I needed help. Yes, I told her, "Yes!" She drove me to the Sugarland Visitors Center just outside Gatlinburg and Connie, bless her heart, turned around and met us there. When I arrived at her house, there was a comfy pair of pajamas, slippers, toiletries, towel and washcloth laid out for me in the guest bedroom. While I was taking a bath, she prepared a dinner of chicken, pasta, corn, and salad. We ate at about 1:00 a.m. It was fabulous! I had run out of fuel and hadn't had a hot meal all day.

I learned that it was Dragonfly's birthday and she had hoped to start her own thru hike on that very day, but it wasn't the right time for her to do it. She was enjoying my trail journal and I think lending me assistance helped her feel part of something bigger, the magnificent Appalachian Trail. Her husband Randy (trail name Sock Monkey) works the night shift as a postal supervisor so I didn't meet him until the next morning. I slept like a baby in their guest room's comfy bed. I took a zero the next day and spent two nights in Dragonfly and Sock

Monkey's home. Meeting with them and staying in their home was an important step in my journey to wholeness and healing and in restoring my faith in the goodness of people.

Dragonfly dropped me off at Newfound Gap in the rain, and I hiked a short three miles to the Icewater Spring Shelter that afternoon. To my dismay I woke up in the wee hours the next morning to discover a new dusting of snow on the ground that grew into a few feet of snow by mid-morning. Post holing with bandaged heel blisters is really painful. Fortunately, I had arranged for a friend of a friend to pick me up three days later at Davenport Gap.

I stayed with Cathy in her comfy mountain cabin in Waynesville, North Carolina for three nights and had time to heal my blistered feet and make a great new friend. I visited Cathy in Waynesville the summer after my hike and enjoyed the place so much I decided I wanted to live there. I applied for a job and got one and moved to Western North Carolina in the Fall of 2015, about a year after completing my hike. I now live a mile off the Blue Ridge Parkway and about 40 minutes from Newfound Gap. Its been a thrill for me to provide trail magic to some 2016 thru hikers at Newfound Gap, Stecoah Gap, Hot Springs, and Max Patch. How awesome is that!

Many AT hikers talk about the "Virginia Blues" since crossing state lines is a tangible measure of forward progress through the 14 states the trail crosses. 500 miles of the trail are through Virginia, almost one-quarter of the entire AT. I spent the end of April, all of May, and the beginning of June of 2014 hiking in Virginia, about six and a half weeks. That's a long time in one state.

About 150 miles into Virginia, I stopped for a rest at Woods Hole Hostel. There's an expression I love: "The great arises out of small things that are honored and cared for." Neville and Michael have honored and cared for countless small things at their hiker haven: homegrown produce, hand thrown pottery, home-made soap, wood burning water heater for the outdoor shower, three cats and a dog, beautiful wooden floors and furnishings in the house, and wholesome meals reverently prepared and eaten in a spirit of community. Before dinner, and then again before breakfast, Neville and Michael ask their guests to form a circle, hold hands, and state their names and what they are most grateful for. This mealtime ritual makes the food they serve nourish more than the body.

Neville and Michael are licensed massage therapists. After a restful night in the barn bunkroom, I booked a restorative, healing massage with Neville. She knows just where a hiker gets knotted up and worked wonders on my shoulder blades, hips, legs, and feet, using essential oils and hot towels to enhance the experience. It was one of the best massages I've ever had and reasonably priced.

Wishing to pass on the healing vibes and wonderful feeling I was experiencing, I paid for an extra massage and asked Neville to give it to another hiker who she thought might need one but couldn't afford it. As I was relaxing on the couch before getting back on the trail headed for Perrisburg, the phone rang and Neville said it was for me. It was Funnybone, the main author and organizer of this book. I hadn't met Funnybone in person, but we'd corresponded through my trail journal. He told Neville that he wanted to pick up the tab for my stay at Woods Hole. I had already paid my bill, so I asked Neville to pay that kindness forward, as well. Kindness and peace is infectious, and both Funnybone and I were on the same page, wishing to share the Woods Hole Hostel experience with others. It's a very special place where greatness arises out of Michael and Neville's honorable caring for small details that truly make a hiker's day.

I hit a low point midway through Virginia just before my friend Sue (aka Upstream) arrived from San Diego to hike with me for a week. Fortunately, friends I met through trail journals, Sandie and Craig, aka Dogmother and Dogfather, stepped up and invited me to stay in their spacious guest room in Salem, Virginia for a few days. Sandie parked her car and met me on the trail and we hiked together for the last mile or so to VA 42. We stopped at Kroger and I bought lots of fresh fruit, juice, some trail resupply, and ice cream. Sandie dropped me off at her house and I took a long soak in her sunken tub while she went to a different store to get food for a barbeque. The bath was amazing! Sometimes a shower just doesn't do the trick to get really clean. I was able to soak long enough to get rid of the sticky residue on my feet from taping blisters, and felt more relaxed and clean than I had in a while.

Sandie's husband Craig got home from work and we had an amazing dinner of barbecued steaks, baked potatoes, corn on the cob, and salad with avocados and juicy tomatoes. Oh my goodness! I enjoyed that meal so much and felt nourished and satisfied. We planned out the next few days of slack packing for me with some hiking with Sandie and Craig over the weekend.

While staying at Sandie and Craig's, my cousin Tom posted a quote on Facebook by Mark Twain that I love: "20 years from now, you will be more disappointed by the things you didn't do than by the ones you did. So throw off the bowlines. Sail away from the safe harbor. Catch the trade winds in your sails. Explore. Dream. Discover."

The kiosk at the entrance to the trailhead at the Mountain Lake Wilderness area, which I passed through shortly before meeting Sandie and Craig, included a quote by Henry David Thoreau: "I went to the woods because I wished to live life deliberately, to front only the essential facts of life . . . and not, when I came to die, discover that I had not lived."

It was good to be reminded why I decided to do an AT thru hike. Sometimes it's hard to remember the big picture when one is buried in the minutia of day-to-day survival. What am I going to eat? Where am I going to sleep? How far do I want to push myself today? The grace given from the trail while staying with my new friends in Salem, Virginia reinvigorated the sense of purpose I had to begin my hike. Once again, this gift came at a perfect time for me.

After that wonderful respite hiking packless, eating real food and sleeping in a bed with my new dog pal, Daisy, my trek though Virginia was also softened by the company of my good friend Sue. She told her friends about the trail name phenomena and they chose her trail name, Upstream, because she loves to swim and is a positive soul. We had shared some hiking adventures before; the most recent a Peruvian trek to Choquequirao, which was a blast. So I knew Upstream and I would have a great week on the AT.

Hiking with Sue was fabulous and we received some great trail magic from my BFF's little sister, Margaret. She met us at a road crossing near the Peaks of Otter and brought us to a laundromat. We had lunch together and then she dropped us off at the Peaks of Otter Lodge.

Another Virginia trail angel who helped us out considerably is Oma, who owns Three Spring Hostel in Buena Vista. What a delightful, nurturing woman who knows how to care for hikers' every need! After a restful night sleeping on clean linens in her bunkroom, we woke up to a recording of the Rascals singing "It's a Beautiful Morning" and the smell of fresh brewed coffee and a delicious breakfast before getting shuttled back to the trail. The day Sue flew back to San Diego, Oma met me at Hogpen Gap with my pack and brought me a hot cup of coffee with cream, just how I like it. What a sweetheart!

So, I was prevented from getting too blue in Virginia by the kindness of strangers and the company of my good friend Upstream. After saying goodbye to Upstream, I was eating dinner with some other hikers at a shelter just south of Spy Rock. A young female hiker said, "Are you Left Turn? I want to thank you for the massage I got at Woods Hole Hostel!" Wow, that really made my day. I hadn't expected to meet the recipient of the massage I donated, and it was a treat to hear first-hand how much it helped her.

Just before reaching the end of the AT in Virginia, I met a hiker 30 years younger than me named Jackie, trail name Scout, at the Bears Den Hostel. We hiked together through the end of Virginia's "roller coaster", past the big rock with "1,000" painted on it, a true "milestone", and into Harpers Ferry to get our pictures taken and be logged in the record book as 2014 thru hikers 470 (Scout) and 471 (me). Scout hiked ahead of me most of the day but we met up for a lunch break and then again after crossing Keys Gap and stayed together for the last six miles into Harpers Ferry.

It was another beautiful day on the AT, warm and sunny with the trees protecting us from getting over-heated. Talking to Scout helped redirect my mind from wallowing in negative thoughts, which were proving difficult for me to shake, about past relationships, lost opportunities, and stuff I had no power to change but couldn't quite let go. Human company helped when my mind wanted to play in the old groove of a record that had been played way too many times. Scout was a very interesting and positive 23 year-old. I was motivated to hike fast enough to keep up with her into West Virginia. The day before reaching the

ATC Offices in Harpers Ferry, known as the psychological half-way point for thru hikers, I was thinking, "Man, I've walked 1,000 miles and I have to do this much walking again, over rougher terrain to reach the end!? Hmmm." My journal entry on June 6:

> I was a little down when I got here, feeling a slight temptation to call it quits when I reach Harpers, but the good company, good food and comfortable accommodations reinvigorated my resolve to be a finisher. I love to start things. I'm historically a lousy finisher. One of my goals for this hike is to change that trait in myself. Finishing this very difficult hike will give me the confidence to see other projects through. I'm percolating ideas for a business I hope to start when I get home. Part of me wants to say I've learned and grown enough from this hike and it would be ok to stop now and go home to work on starting up my business. But, I think it's pretty important that I stay disciplined and finish what I've started here. Time to pull out my three lists...

While preparing for my hike, I read *Appalachian Trials: A Psychological and Emotional Guide To Thru hike the Appalachian Trail*, by Zach Davis. Zach's book suggests that mental preparation factors in higher than fitness or the right gear in determining whether an aspiring thru hiker is able to complete the journey from Springer Mountain, Georgia to Mt. Katahdin, Maine. He suggests making three lists and referring to them on the days one is tempted to quit. I published my lists in one of my pre-hike blog posts back in February:

1st List: Why am I doing this?

I am Thru-Hiking the Appalachian trail because...
1. I aspire to be more self-reliant.

2. It is something I've dreamed of doing "some day" -
Today is a good day!

3. I love the outdoors and I love to walk - Carpe diem!

4. I'd like to steer my life in a more fulfilling, creative,
courageous direction.

5. I'd like to attract healthier relationships.

6. I will never be younger. Aging happens.

7. I love backpacking and I have some good gear I
purchased when I worked at REI a few years ago.

8. I desire a closer relationship with God and walking the
AT will relieve me from my distracted busy-ness.

9. Completing a 2,100+ mile hike will instill me with the
confidence I need to take brave steps in my career and
future.

10. The very idea of hiking the Appalachian Trail excites
me more than anything else I can think of . There is
nothing I would like to do more.

10. I've been depressed/disillusioned since my
divorce/relocation and this hike will reset my psyche.

2nd List: What benefits will I receive from completing the
trail?

When I Successfully Thru hike the Appalachian Trail, I
will...

1. Be in incredible physical/mental/spiritual condition,
ready to face life's challenges and obstacles.

2. Feel self-reliant, empowered and confident.

3. Have opportunities to share my story of strength with
other women in transition.

4. Feel closer to nature and to God.

5. Gain confidence in following through on creative ideas.

6. Have had a lot of time and space to process unresolved
grief.

7. Gain clarity about where I'd like to steer the second half
of my life.

8. Be better able to fulfill my purpose (to experience and
share joy).

3rd List: Negative Consequences associated with quitting the trail

If I give up on the Appalachian Trail, I will...
1. Disappoint myself.
2. Appear weak-willed and foolish.
3. Not receive all the benefits I just listed.
4. Lose the burst of confidence that planning this adventure has given me.
5. Give myself the message that I'm not a finisher.
6. Fail to grow/transform my current funky state.

When I woke up the morning after casually mentioning quitting in my trail journal, I received an outpouring of heartfelt messages from my trail journal readers. Here are a few that can still make me tear up: "Congratulations on Making it to Harper's Ferry! Arggggh, you just have the Virginia Blues, take a couple of Zero Days and hop the train into DC (there is a hostel in downtown DC) and see the sights. Don't force yourself into a hike you aren't enjoying, but don't quit while you are tired. ... There are so many wonderful experiences and places ahead of you - you wouldn't believe but all of the amazing experiences that you have had is just the warm up and preview of what is to come." Jun 6 2014 11:35AM, From Sit A Bit, W Kennebunk.

And another message from Melanie of Pennsylvania: "Jackie, You CAN do this! ... In '88 I quit my attempted thru hike at Franconia Notch, NH. I was severely undernourished and emotionally/physically exhausted. I went home. A week later I regretted my decision. And even though I went on to finish up the AT in the following years, I always regretted quitting my thru hike. The trail...the experience...it becomes a part of you that only long-distance hikers can fully understand. Whatever you decide, it has to be right for you. You have come a long way from where you started in many ways. (I'm sure this journey has been much more difficult than you have shared in you public journal.) You have such a positive spirit, send out positive energy and have inspired many of your readers (me included). I am pulling for you Jackie. I think it's safe to say WE, your devoted readers, are pulling for you. Be well Jackie."

Allen from Lebanon, Tennessee wrote: "Jackie, Don't Quit!!!! You have an incredible following. I need you to keep going! You have the awesome privilege to accomplish a dream that thousands of dreamers following you never will. You quit, we quit! Roll on LT!"

A reader from Gloucester, Virginia wrote: "No quitting now, girl! These are the moments that test your mettle. Think your mind's stuck in a groove of remorse and "what ifs" on the introspective days? Just think of the remorse you'll feel back at home if you quit your dream! I am a San Diego gal born in 1960, too, and plan on following in your footsteps next March. You are an inspiration! Go forth and conquer a day at a time. I will be cheering as you summit Katahdin! :)"

Needless to say, from this point forward I knew I had damn better do my best to finish! My peeps were counting on me! So, walking into West Virginia, I was getting more comfortable with accepting small and large kindnesses from Trail Angels. A trail journal reader named Vicki offered to pick me up at Harpers Ferry and treat me to a meal, shower and room in her home in North Potomac, Maryland. When she met Scout, she invited her to come along and Scout and I shared a queen-size bed in Vicki's guest room. After sleeping in shelters and bunkrooms with a bunch of men, sharing a bed with a female hiker I just met the day before was not a big deal. Vicki's husband Howard was not as into the trail as Vicki but was pretty good-natured about having us crash at his place. We enjoyed a delicious barbecued dinner and conversation. Their home was cozy with two big fluffy cats, some finches and parakeets. It was nice to stay in a home with pets.

Some hikers take part in a three state challenge and hike 40 miles through West Virginia, Maryland, and cross the Pennsylvania border all in the same day. I, on the other hand, languished through Maryland, and slept indoors through most of it, thanks to Vicki and Howard, and another friend I made through trail journals, ZigZag. She slack packed me and invited me into her home in Germantown, Maryland. We even went to see a matinee movie, *A Fault in the Stars*, which I had read after picking up the hardcover book in a shelter. Even though the movie was so-so, the normalcy of sitting in a movie theater halfway through my hike was very meaningful for me. ZigZag completely understood what I needed because she had attempted a thru hike in 2011. She had to leave the trail before completing some parts of New York and Maine. She completed the missed sections later that summer and I met up with ZigZag and her lovely daughter, Laura, farther up the trail in Greenwood Lake, New York. After that, she hiked just a little ahead of me on the trail and texted me advice, about where to stay, eat, and resupply. ZigZag summited Mt. Katahdin four days ahead of me on 9/8/14 and accomplished her goal! Yay!

So, this brings me up to the Mason-Dixon line and my entrance into Pennsylvania, lovingly nicknamed Rocksylvania because of the tectonic shifts that caused the boulders to land on their sides with their pointy edges sticking straight up into the air, twisting hikers' ankles, tearing Achilles tendon, and wearing out our shoes. Shortly after reaching the Half Gallon Challenge at Pine Grove Furnace State Park, I met the amazing trail angel who assisted me through Pennsylvania, Iceman (real name: David). He got the name Iceman because he had a bag of ice to assist a hiker who had broken her arm, not for an icy disposition.

Iceman had been reading my journal and the first time I met him, he was parked at a road crossing in Pennsylvania, waiting for me. He stuck out his hand to shake mine and said, "Hi, I'm Iceman." I had no idea how instrumental he would be in my being able to accomplish a successful thru hike in good health and relatively good spirits. When I met him, I was, again, pretty tired out. I thought about quitting at Harpers Ferry. The trail magic that Vicki and Zig Zag provided in Maryland helped me tremendously, but I was worried about getting through rocky Pennsylvania. I sat at the Half Gallon Challenge in Pine Grove Furnace State Park, barely able to eat a pint, wondering, "What am I doing here with all these kids? I can't eat a half gallon of ice cream, and I can't hike as fast as them. I'm never going to finish this dang hike. I'm tired. I want to go home."

After Iceman stuck out his hand and introduced himself, he pulled a camp chair out of the back of his truck and invited me to sit down and drink a Mountain Dew. A chair is such a luxury on the trail and Mountain Dew is an elixir that revitalizes body and spirit. I had been thinking about switching out my Gregory Diva 60 for a lighter pack. So, I ordered a new Osprey Aura 50 on Amazon.com with my phone from the comfort of Iceman's chair on the side of the trail. He offered to have it delivered to his house and bring it to me when it arrived. Iceman also drove hours out of his way on Father's Day to pick me up on US 11 and drive me into Carlisle, Pennsylvania, where I was to meet my daughter, son-in-law, and granddaughters to spend a zero day together. What a prince! Iceman slack packed me for a total of nine days; four in PA, one in NJ, two in MA, and two in NH.

Another delight that the state of Pennsylvania brought me was hiking with Dallas, another middle-aged woman, an athletic dynamo from Baton Rouge Louisiana. Dallas became my hiking partner through the next seven states. We hiked about 670 miles together, splitting the cost of motel rooms, sharing meals, keeping each other company, and laughing our heads off. Dallas told me she was thinking about selling her house in Baton Rouge, although she'd lived in Louisiana her entire life, and moving to Grand Junction, Colorado. Well, by golly, she did just that. I visited her in Grand Junction the summer after the AT, and did some high elevation hikes with her. We've been toying with the idea of hiking the Colorado Trail together.

Iceman surprised Dallas and me by calling when we were hiking in Massachusetts. He slack packed us for Dallas's birthday weekend. We were so surprised that he was willing to drive such a great distance just to help us out for the weekend. Dallas and I had so much fun standing at the top of Mt. Greylock with Iceman, watching the hang gliders and paragliders sail through the sky. Dallas was thrilled with the Dunkin Donuts Iceman brought her for her birthday breakfast. I ate quite a bit of the seven-layer salad that Iceman's wife, Barb, made for him to take along. Something green!

Just when we thought the story of Iceman and his unbelievable trail magic could not get any better, he surprised us one more time when he and Barb drove up to New Hampshire to meet Dallas and me. This time Barb did the driving and Iceman, Dallas, and I were able to hike with light daypacks through the challenging sections between NH 25, Kinsman Notch, and Franconia Notch. Those were not easy hiking days, but they would have been much more difficult had Dallas and I been carrying our heavy backpacks. Iceman and Barb's well-timed, thoughtful, selfless assistance worked wonders to heal my feet, re-energize me, and warm my heart. Damn yes, people are good! Iceman section hiked the AT over a number of years and completed his years-long goal of summiting Mt. Katahdin on August 7, 2016. Woot woot!

In New York, I received a delightful surprise when my old friend Jennifer from San Diego contacted me to let me know she was going to be near the AT on vacation and that she wanted to provide some trail magic! She met Dallas and me at the Native Landscapes Garden Center in New York and treated us to a picnic lunch. Small comforts like seeing an old friend are magnified a thousand fold when one is walking into unfamiliar territory each day. Jennifer's kind thoughtfulness will never be forgotten.

Farther up the trail in New York, near Canopus Lake, Plantman and Pat invited Dallas and me to stay in their home for two nights during raging thunderstorms. We ate some delicious vegan meals, enjoyed a day of slack packing, and got to meet another hiker, Slim, from Boston, who was also treated to food and a bed courtesy of our vegan hosts. Plantman finished up his missed section of the trail in 2014. It warmed my heart to see his picture with the sign on Katahdin.

Farther up the trail, Dallas and I faced rainy weather, vicious biting insects, and chronic foot pain, as well as flooded muddy trails at the CT/MA border. Dallas took a bad fall which left her hanging upside down from a tree – really scary! We were ready to spend some time indoors.

Shortly after crossing over Mt. Everet, Father Bruce drove up with his beautiful German Shepherd Dog, Malakai, just as we hit the highway. Father Bruce is the pastor at Our Lady of the Valley Catholic Church in Sheffield. He arranged for us to stay at his next-door neighbor's house because, "She takes in hikers." He said he would have let us stay at the rectory but it's a small town and people would talk. :-)

Dallas and I got cleaned up, did our laundry and enjoyed sitting with Father Bruce and Malakai on his comfy, breezy screened-in front porch. We went to dinner at Baba Loui's in Great Barrington. While we were waiting for our table, Dallas and I bought a variety of energy snacks at Robin's Candy Shop, which carried about 50 different kinds of licorice from all over the world. Quite a place! It was nice to catch up with Father Bruce. He and I had met in Poland a few years ago at a retreat led by Bernie Glassman at the Auschwitz/Berkinau concentration camps. We've stayed in touch on Facebook, and back in February, when he read about my planned AT hike, he let me know he lives close to the trail and offered trail magic when I got up to his neck of the woods. That was five months earlier when it seemed such a remote possibility that I'd sit down to dinner with Father Bruce in Great Barrington MA!

So, when we got back from dinner, Dallas and I walked next door with our packs only to find a hiker we knew named Belch sitting in the living room. "What are you doing here and how did you find this place?" I asked him. He said, "It's in the AWOL guide." Turns out Father Bruce's church and rectory are next door to Jess Treat's house, the Sheffield hiker hangout. Dallas and I stayed in an adorable room on the third floor with attic ceilings and really comfy beds, courtesy of Father Bruce.

Dallas and I hiked through Vermont and the Presidential Mountains of New Hampshire together and parted ways at Pinkham Notch where Dallas stayed an extra day at the White Mountains Lodge and Hostel to get over an intestinal bug. I wouldn't be lonely too long because I soon met my amazing trail angel and friend, Cherri Crocket. Cherri was writing outdoors articles for the

Rumford Falls Times and Bethel Citizen in Maine and she followed my hike on Trail Journals from start to finish. Her wonderful husband, Andy, is a real live lumberjack and they live in a picturesque cabin they built themselves. These are my storybook friends who opened their home and their hearts to me when I crossed the NH/ ME border. Some of the toughest hiking on the AT is in the state of Maine and my load was lightened considerably by Cherri and Andy, who set aside time in their schedules to shuttle me on and off the trail with a daypack, cook amazing meals for me (including a steak and lobster feast), and show me some delightful sights in their neck of the woods.

After saying goodbye to Cherri around the time I reached Rangeley, Maine, we made a plan to meet up two weeks later in Baxter State Park so we could summit Mt. K together with the Warrior Hikers on September 12.

I had three incidents in Maine that tested my confidence and determination. I took a tumble down a steep boulder field above the South Branch of the Carrabassett River. In the sweaty, hot late afternoon, I grabbed onto a tree limb that was not stable and when it broke, I fell pretty hard down some boulders and jammed my left ring finger. I had a little meltdown on the side of the trail, my finger hurt something fierce, and the fall really scared me. As I was sitting there crying a SOBO hiker named Buck Eye came along and cheered me up. Then, I took another fall on a slippery rock just north of Caratunk and tore a huge chunk of skin off my left palm. Funnybone had been sending me encouraging messages in my trail journal from time to time, and after this fall, he gave me some sound medical advice for caring for my open wound. I had some iodine tablets in my pack which were my backup for emergency water purification, and he told me to mix some with water to rinse the wound with the iodine solution. It worked!

A few other incidents in Maine shook my confidence: getting lost in the woods trying to find a suitable flat spot for my tent and then losing the trail for about 20 minutes. Very scary, especially now after learning what happened to Inchworm, an experienced hiker who died of starvation just a half mile off the trail because she got lost. My other knuckle-headed move was to attempt to night hike to get to Abol Bridge. I believed a south-bounder who told me that section of the trail was a flat road walk. Wrong! It was mountainous, and dangerously close to a roaring fast stream. I hiked until 11 that night, then put up my tent on a lumpy mound of moss before passing out from exhaustion.

I described these incidents in my trail journal, and my loyal followers were right there for me with just the right messages to lift my spirits. One reader from New Jersey named Dan wrote me almost every day and this message in particular helped me get through my most difficult days in Maine: "'There is in every true woman's heart a spark of heavenly fire, which lies dormant in the broad daylight of prosperity; but which kindles up, and beams and blazes in the dark hour of adversity.' Washington Irving, The Sketch Book, 1820."

After getting a ride across the Kennebec River on "Hillbilly Dave's" ferry, I checked into the historic Sterling Inn, a fantastic place to attend to hiker needs (laundry, shower, resupply). It is also an exceptional place to be surrounded in historical charm and elegance that the owner Eric makes great efforts to maintain (sunken claw foot tub, historical photos, and mementos, adorable common area and guest rooms). I was happy to indulge in a double bed with lots of pillows, a nice soak, and the "grill your own burger" lunch special. As spoken by the late Dorothy Parker, author, humorist, poet, and wit, "Take care of the luxuries, and the necessities will take care of themselves."

After a restful night at the Sterling Inn, I was awakened at 6:30 a.m. by a knock on my door, letting me know I had a phone call. It was Mailar on the phone, an '05 AT hiker who had been following my journal; he had emailed me a few weeks prior, offering a slack and stay through the first 30 miles of the 100 mile wilderness. I tried to call him once I got to Caratunk, but had a wrong number, so I sent him an email asking him to call me at the Sterling Inn. I was so happy to hear from him! We made arrangements to meet at Bald Mountain the following day. I wasn't quite sure what to think of Mailar, so I asked him to please bring his wife Judy, and also asked a young hiker named Wiki to accompany me and share Mailar's trail magic. Wiki was a 25 year-old from Virginia who had plans to summit Mt Katahdin on 9/12, which was also my goal. An ambitious goal for me, but now with Mailar's help, it looked like 9/12 could be doable.

Staying with Mailar and Judy in their hunting cabin in Shirley, Maine was amazing! They cooked us wonderful food and Mailar's exuberance for the trail was energizing. The comforts Mailar and Judy offered me were priceless and I am so grateful for the loving care they provided. Just about everything they served at their table was grown or made by them, and it was a treat to sit down in the evenings and learn about their life in Maine.

After slack packing the first 30 miles of the 100 mile wilderness with Mailar's help, I completed the last 70 miles in four days and reached Abol Bridge on 9/10, two days before my summit date. It poured rain on 9/11, the day I walked through Baxter State Park chatting with Little Spoon, a young hiker from North Carolina. I got very lucky and had fantastic weather for summiting Mt. K with my new best friend, Cherri Crocket, and the Warrior Hikers.

After backpacking through a sampling of weather from three seasons, including blizzards, rain, crazy ruthless biting insects, and many, many days of beautiful sunshine and lollipops, I completed my 2,185 mile hike on September 12 at the northern terminus of the AT, Mt. Katahdin, Maine. While I was on the trail, I developed an immense appreciation for my place in the natural world. Nature is exquisitely beautiful, and I am part of what makes it so! My instincts and survival abilities grew as well as my muscles, and I made some fantastic friends while building self-esteem, confidence and faith to carry me through the second half of my life.

How, you may ask, does one hiker manage to meet so many incredible trail angels during her thru hike? Most of these trail angels found me on Trail Journals. I'm not sure I could have made it to Katahdin without their help. I know it would not have been as much fun! If I could only share one piece of advice with future AT hikers, it's this: Write an online trail journal and update it frequently with honest, sincere reflections about your hike. You will not be disappointed by the remarkable kindness of strangers who connect with you through your writing. I received over 800 messages from people following my hike. Probably the most touching was from a clinical psychologist in Tulsa, Oklahoma, who told me he had a client who was a young, athletic woman who engaged in negative self-talk. Reading my trail journal became part of her therapy. Wow! I am humbled and grateful that sharing my experience might help others.

You may wonder how this AT experience might have changed me. Did I accomplish what I had hoped to achieve as set out on my three lists? Heck yes! Thru-hiking the AT transformed my floundering self-doubt into a confidence in my inner compass and instincts, a feeling that I can survive in this world. If the economy collapses, I know how to live with just 30 lbs of stuff and I can even carry it on my back in all sorts of weather, right? My faith in humanity has also been restored, largely due to the kindness and compassion of strangers.

I now live in the mountains of Western North Carolina and hike almost every weekend. I've recently been asked to help out in backpacking basics clinics and I am sharing my story of strength with other women, just as I had envisioned and hoped before I started my walk. I work in a pottery studio/shop. My job is not stressful. I work hard and play hard, but I also had my bathroom remodeled and have an amazingly deep and luxurious clawfoot tub. Honestly, I'm genuinely happy for that brick wall I hit, because it motivated me to pull out the stops and take the greatest adventure of my life.

My faith in humanity, in myself, and in God has been restored. I am whole. Living the dream, fulfilling my life purpose to experience and share joy. I hope my story may encourage you to do something scary. Go for it! It's so worth it!

Jim Dashiell

Cherie Crockett and Jackie aka Left Turn

~ ~ ~

I grew up in Chicago, attended the same elementary
school as my mother and grandmother, and even had one
of the same nuns. There was no PE offered. I didn't play a
team sport or run a mile until we moved to the suburbs
and I began attending public school at 13 years of age.
Needless to say, I wasn't much of an athlete in young
adulthood. So, I'm in my mid-30s, living in
California, sitting on the couch smoking a cigarette when
the phone rang and the guy on the other end asked when
I'd like to come in to start my one week free membership
at 24-Hour Fitness. My preteen kids heard me say, "I
signed up for a gym?" They were snorting and snickering
-- they had signed me up as a joke. The joke was on them
because I took that free trial membership and liked it; so I
signed up and started doing step aerobics. I enjoyed the
exhilaration and endorphin rush from working out, then

128

I'd enjoy a cigarette in the gym parking lot right after class. Eventually, I quit smoking and engaged in a few more athletic activities like biking and yoga.

After the sudden death of my 17-year old son, David, I began hiking for mental health with my friend Jo Anne, who I met in a Hospice grief group for mothers. Jo Anne and I met at least twice a week to hike up Cowles Mountain in San Diego. We talked about our children who had passed, my David and her Cassandra, as our friendship grew. We took a few road trips and explored trails in Yosemite and Zion National Parks.

In 2011, I got a job at REI and purchased some backpacking gear and started to do a few overnights. My friends Debbie and Jeff taught me how to pack my backpack, set up my tent, set up a kitchen and cook at camp. We took some awesome trips in the Eastern Sierras.

Then, in 2012, I got a divorce and moved to Florida to be close to my daughter and her family. My backpack started to get a moldy smell and I was missing the mountains. In late 2013 a series of events began to chip away at my positive outlook on life. A combination of insight, opportunity, and space opened up in my life, making it possible to attempt a 2014 AT thru hike. I went for it, not really believing I could do it. I surprised myself. This is my story.

Chapter 8 Woods Hole Hostel: The History, The Story and The Magic
Neville Harris

Who are we? We are Neville and Michael. We run Woods Hole Hostel along the Appalachian Trail in Southwest Virginia. The hostel, like the hikers, has it's own wonderful story. Michael and I have incorporated our own dreams and ideals into the vision that my grandparents started back in the 1980's. They opened a hostel along the Appalachian Trail out of a desire to meet all of the unique individuals who hike through their "back yard" – the National Forest.

First, it seems only right to mention how Jim (Funny Bone) met Michael and me. You and some friends had decided to do some "shake down hikes". You all decided to use our hostel as a base for your hike. I re-call that you and I hit it off right away because my father was also an orthopedic (bone doctor) surgeon. We seemed to have a beautiful kinship through my dad.

You came with the idea that you would do a thru hike in the Spring. Honestly, I will be honest, I never thought you would make it. Anybody can make it with enough determination, but I was super surprised to see you walk onto our porch that Spring with the herd of thru hikers. You came to say hello, have a smoothie, but had already decided that you could not stay the night. You knew that in order to do a thru hike you would have to keep a certain pace for your entire hike and this would not allow you the luxury of staying with us.

That's ok, we are still grateful we got to meet you.

A few years later you asked us to write about hostel life, perhaps our history, what motivates us, and more. Where shall we begin? It always seems best to begin with our history. I like to think that we have a beautiful history; a deeply layered love story.

It begins in 1880, if not before. A man named Stoney Holiday came up Sugar Run Valley, up to uncharted land, and found a slice of land where he could build a Chestnut Log Cabin. Those were the days when Chestnut Trees were in plentitude. I imagine that he literally harvested the trees within feet of where the cabin stands today. After only a short spell of living here, he and is wife found it too cold and too remote, and so they departed. Another family came along, and another, until finally the cabin was left to be abandoned. It was simply too remote a place for human life.

In 1939, on Christmas Day, my grandparents went looking for a place to live in Sugar Run Valley where my grandfather could do a study on Elk. He was a graduate student at Virginia Tech and had been offered a stipend to study Elk that were in the region. He and my grandmother went searching for a place to live. First, they were pointed toward a tenant cabin, but my grandmother did not like the wall paper. They asked if there was anything else and my grandparents were directed toward the dilapidated 1880's Chestnut log cabin and 100 acres, now, up against the National Forest and Appalachian Trail. In those days the 30,000 acres behind us were privately owned land. The land itself was barren. All the trees had been cleared for farming. My grandparents fell in love with the old cabin even though it had no electricity, no running water, and holes in the seams of the cabin. They rented the old cabin, which was more like a barn, for the year and he studied the Elk to see if he could find a way for the Elk and farmers to co-habitat. Elk were destroying farmer's crops because they had nothing else to eat. My grandparents stayed for the year attempting to learn as much as they could. By the end of the year, the stipend my grandfather received ended and he moved on to a career in Environment.

After a few years, the neighbors learned the land my grandparents had occupied was for sale. It was cheaper for the neighbor to buy the 100 acres and old cabin than build a fence -- $300! The neighbor called my granddad and together they bought the land. Time passed, and my grandfather got another call -- the neighbor was ready to sell.

When my grandfather retired in1981, he, my grandmother, my sister, and myself began making the trek north from Georgia to visit the Chestnut Log cabin as a summer home. It was not long before my grandfather recognized that the Appalachian Trail sat right next to the property on the National Forest and he began dreaming about inviting hikers to the cabin, sort of like the hostels he had seen in Europe.

You see, in those days, there were not too many hostels on the Appalachian Trail. As a matter of fact, there were only a few. My grandparents contacted the Appalachian Trail Conservancy (ATC). They wanted to know what they would have to do in order to open a hostel. The ATC suggested they could try, but there may be little interest by hikers. A shelter was only two miles away and town was only 10 miles. My grandmother, also, did not like the idea. She did not want to invite those "smelly hikers" into her home. But my grandfather was persuasive, and she conceded. So, in 1986 my grandparents opened their log structure home, they came to call a bunkhouse, to hikers. In the first year, they only had a handful of hikers – 12, I think. In 1987, they opened their doors again. In the second season, my grandfather passed away very suddenly of a heart attack. It was a terrible surprise. I always like to consider it a blessing that he passed away at the cabin. It put him closer to heaven and my grandmother closer to this home.

At first, she wasn't certain if she would continue running the hostel, but as she gained support from the ATC and her family, she found herself annually trekking to the hostel from her home in Georgia for the hikers. She felt her service was greatly appreciated by all the hikers. She continued running the hostel for 22 years, up until her own death in 2007.

In the Spring of 2007, the week before she was to head to the cabin for hiker season, she learned that she had cancer. Her response, "Those Doctors had better get their acts together. I've got to get to the cabin for Hiker Season." My mom managed to persuade my grandmother to stay home a little longer so that she could get proper care before heading off. She was given 6 months to live and that is how long she lived and man did she live!

So, now, I guess, I best back up a bit and explain how Michael and I fit into this whole story. Well, like I mentioned before, my sister and I had been coming to the cabin since around 1981. That would put me around the age of three years old. I still have fresh, raw memories of the first summers we spent at the cabin. I remember the smell of the musky cabin, the pinch of the aluminum chairs against my skin, taking baths in the kitchen sink, and playing in the creek.

Each summer my parents would help my sister and I find a way to come visit my grandparents for practically the whole summer. This impacted me greatly and I continued visiting beyond college days. I graduated college in 2000. The year after college, a childhood friend and I decided to embark on an adventure before heading for real jobs. We decided we wanted to be like Henry David Thoreau. This was our opportunity to go live in the woods. We did not have grand thoughts of hiking the Appalachian Trail like others did. Instead, we had the grand idea of going to live at my grandparents cabin and hostel for the Fall of 2000. It was the first time ever that we had opened the hostel for South Bounders and I absolutely loved it. Suddenly, I found myself saying, "Wow, I really would like to live here one day." But that felt like a far off reality. Each summer, I continued to make an excuse to come visit my grandmother for however long I could. I would come between jobs or school -- and that is how I met Michael.

In the summer of 2005, I decided to head to the cabin for about 10 days to visit my grandmother. Now, just to give you a little personal history ... When I met Michael, I had already met a lot of Appalachian Hikers and been attracted to all sorts. But, I also like to add that I had never hmmhmm... "hooked up" and I certainly did not 'hook up' with Michael. But meeting him was just a tad different. When I met him, I told myself, "I'm not going to fall for this hiker!" So I made every effort not to, even though, to this day, I think I did! Just before he left I decided to give him my email address. One random day, that fall, I heard from him. We began to email and slowly became great friends. Our dreams coincided beautifully and our friendship grew into a bond. We both talked of organic gardening, community lifestyle, and more. We had no idea what was in store.

As we fell in love, I kept visiting my grandmother at the cabin. Over the years, she and I had formed a powerful bond, something very unique and rare. I don't know too many granddaughters that "hang out" with their grandparents. But that seemed to be me. So, when my grandmother passed away, she had already made comments to hikers of how, "You had better come to the cabin now, because when I die, Neville's going to run the place and every thing's going to change!" Well, I guess we did end up changing a lot of things. But we like to think it was all for the better.

Michael and I got the opportunity to move to the cabin in the Spring of 2009. My mom had seen the work Michael and I had done to our own house in Dahlonega, Georgia. On one small acre, he and I had built a garden, were raising bees, had a small flock of chickens, and were even in the process of building a Sauna in the back yard. She felt confidant that he and I could handle "living in the woods". We left the small town of Dahlonega (ironically, the southern summit of the Appalachian Trail) in the Spring of 2009. We left not knowing how we would make a living or a go at running the hostel. We just knew that it lined up with our life goals and it was a chance worth taking. We lived out our first year and many more years facing challenges we never expected to face, but we've also learned lessons, gained experiences, and fallen more and more in love with our lifestyle.

Fortunately, Michael had been a thru hiker. And I knew the thru hiker lingo (words like "blue blaze", "slack pack", or "zero" can seem like a foreign language to the outsider) but I knew this "language." However, I did not have the experience of thru hiking the trail. Michael was able to provide an insight that greatly benefited the services we would come to provide. Where my grandmother had only offered a free bed and a breakfast for $3.50 (true!!), we began to offer shuttles, laundry, a "real" hot shower, indoor rooms, massage therapy, free yoga, communal meals, and more. We were carving our way so that we could make a life of being and living on the mountain. Each year since we have been here, it seems that we are always improving upon ourselves. We have never had a "bad season", but we seem to learn a lot from each season and want to make the next season even better.

Our year is broken down into seasons, just like most everyone else. We have the hustle bustle of thru hiker season in the spring, the lazy days of the summer, firewood cutting in the fall, along with the small groups of southbound hikers, and reflecting in the winter. It is not a lifestyle that would suit everyone, but it seems to suit us. It truly has been a gift to serve hikers. Doctors get to serve patients, teachers get to serve students, bus drivers get to serve travelers, and we get to server hikers and the occasional person that's looking to get away in the woods.

When asked by a journalist, or by our many guests, what is the hardest thing about running a hostel and living in such isolation at certain times of the year, I can tell that the questioner expects I'll say "chopping wood" or "the daily grind of making meals for so many hikers". I have to say I may be overly honest. There are many aspects that are hard. Everyone likes to hear about the hard days. Rarely do we speak of the good ones. There certainly are hard days, when we over exert ourselves with all the tasks

Michael and I have created, the guest that somehow "rubs us wrong", and the relentless duty of keeping the house clean. All of these jobs are daunting and exhausting, but in truth the hardest is relating to my business partner, my lover, my best friend, my co-coordinator in life, my husband. When we get along everything seems so seamless and perfect. When we don't get along everything halts with a vengeance and one feels like quitting as fast as one can! But at the end of the day, the year, the week, or the month, we continue to persevere, pursue, and try. For what reason, we don't really know. Perhaps it is our love or perhaps dedication. We both seem to bring our own gifts to the table. And combined we seem to make a perfect pair.

I've never hiked the Appalachian Trail, but I have grown up along the Appalachian Trail since I was three. I've watched how the small kind deeds of strangers can impact hikers and keep them motivated toward their goal. We have had hikers over the 30 plus years walk in ready to quit and leave feeling restored from something they experienced. Perhaps, all they needed was rest, a good meal, or the kindness of a stranger. We will never really know. But in a way, they leave here with our hearts as full as their own. Hikers keep us coming back and we keep hikers going. It is a wonderful exchange of energy.

We do meet our fair share of "bad hikers" as well. They've always existed, and probably always will exist. Back in the 1980's there was a hiker named Duffle Bag Tim. Nobody wanted him to show up at their shelter or at their hostel. Everybody new that if he showed up, all he was going to do would be beg and mooch. I remember the day he showed up here. We were lucky that right when he showed up a truck was heading into town. He bummed a ride and that was all we saw of him. Personally, and maybe not like my grandmother, I think these hikers may be my favorite characters. I enjoy how we get to impact their lives and maybe turn them in the right direction. They are the folks we would never get to share a meal with or share a conversation. Somehow along the Appalachian Trail, we grow together under unusual circumstances.

Just like along the trail, when a hiker pulls up to a shelter and hasn't had anyone to talk to in hours, days, or sometimes even weeks, that same hiker pulls into our hostel, worn out by the trail. We are able to offer them a comfortable mattress, a hot shower, and a hot meal at whatever cost thanks to the "Broke Hiker Jar". The Broke Hiker Jar is a tip jar we created for those that would like to provide trail magic to hikers. They leave money in the jar and we are able to provide them with a service that perhaps they had not budgeted or did not have the funds for. The hiker leaves not only feeling grateful to us, but also grateful for the greater community that has provided. The jar has allowed us to give whole-heartedly, while at the same time, not taking the burden entirely on our own backs. It has allowed us to serve hikers, who otherwise we may not have met.

The jar is much like Trail Magic and just like hikers experience Trail Magic along the Trail. We experience Trail Magic here. We experience the magic of hikers willing to help us operate, build, and repair. When I was a little girl, my grandmother had the misfortune of the oven breaking. She told the hikers she couldn't make breakfast because the oven was broken. One of the hikers spoke up and said he was an electrician! He was able to re-wire her oven and they were able to have breakfast the next morning. My grandmother quickly learned that any time she needed something fixed all she would have to say was, "Let's pray for a plumber, an electrician, or carpenter," and within hours they would appear -- literally!

The trail was providing it's own magic for us. Michael and I grew to understand this same concept. Whenever we are in need, we just look at each other and say we better "pray for a helper, extra hands, or anything" and within hours it appears. Hikers seem to think "trail magic" is trail limited, but we've come to understand it's not. It happens in the real world to. The world we all exist in, it's just a matter of looking around and experiencing it.

By operating a hostel we've gotten to meet fascinating people. But we've also gotten to meet our fair share of bad folks as I've said before. We'll get a call, an email, or just here from a hiker, "You better watch out for this ___ hiker." At first, it made us nervous. How should we handle the situation if they were to show up. When the first hiker did show up, we knew just what to do. We didn't tell them they had to leave or they couldn't stay here. Instead, we told them what we had heard with an understanding that they could stay but they had best not repeat these past actions, like stealing, bumming, doing drugs, etc. The hiker is always quick to give their side of the story, which I always found fascinating. I also remind them that it was not our place to discern the truth, but rather to give them a second chance. In giving them a second chance, most of the times they were happy to redeem themselves, but occasionally, they acted out and again. We would pass the message along the trail that either they redeemed themselves and turned out to be a very nice hiker or that there really was something wrong. Either way we never lost too much and always gained more perspective by communicating with them.

There are good days and bad days to operating a hostel. The good days are when a super nice group of folks walk up. They all seem grateful for whatever you have to offer and are ready to serve, even when it is us that is supposed to be serving them. The bad days are, well, there are no bad days. Sure, there are some questionable moments, like when that hiker you "heard about" shows up and you aren't quite certain how they are going to react to what you heard. Or, the hikers decide to break your rules and "party anyway" – but with our location we meet so few of those hikers that we seem to be incredibly blessed by the people that pass our door step. They are enamored by our lifestyle, and we are enamored by their gratitude. Hikers seem to have a purpose, even if all it is, is to walk everyday. They seem to be ready to help and open to experiences. We are so grateful to the hiker community for all they have done and all they do to make Woods Hole what it is today.

~ ~ ~

The author of this next chapter is not a hiker, she's a hostel owner of a very special hostel on the Appalachian Trail. Neville and her husband, Michael, literally fell into hostel ownership. Her grandmother had operated the hostel for 22 years and when her grandmother, Tillie Wood, passed away in 2007, her mother offered them the opportunity. Both said, "Yes!" Well, actually, Neville said "yes" but then had second thoughts and Michael said, "The U-Haul is packed." Both brought their own skills and dreams to the hostel. She had studied to be an artist and school teacher. He had hiked the Appalachian Trail and grew up learning how to garden and be self reliant. The two have become a dynamic pair operating the hostel.

See for yourself. Visit for yourself. Neville's chapter gives you a glimpse of the unique history, the life they live, and the way they operate their hostel and now bed & breakfast.

A Poem **My Katahdin Eve**
Patricia Doyle-Jones   aka Red Robin

I hiked the Appalachian Trail
What on earth did that entail?
A daily walk without fail
All supplies were sent by mail

Springer to Katahdin
Georgia to Maine
Many thought me totally insane.

Mother Nature did not care
That I was planning to be out there
Rain and shine. Wind and cold
Many times she was bold

State by state
Always something great
Peaks and valleys, Streams & springs,
Birds & bears, & other things,
Bridges & bogs w/ boards a floor,
Towns & stores, w/ food to adore

Hostels, homes & huts around
Places of respite for me northbound
Trail magic provided along the way
Always managed to brighten my day

The walk along from Spring to Fall
Awoke in me an earthen call
To appreciate all what I can see
As so much doesn't come easily

Spring bluets & beauties brightened my way
I loved the rhododendrons everyday
The forests of trees from oak to fir

Stood strong & fragrant in the windy stir

Many friends whom I adore
Gladly opened wide their door
Then Trail angels with help galore
Their kindness appreciated all the more

When friends & family
Joined my hike
Those are times
I really liked

An amazing support came from all around,
Notes of encouragement did abound,
All for me that I stay strong
So I could hear Katahdin's song

As I arrive, I stare at your peak
Very austere & blatantly bleak
I thought a joyous sound
I would shriek
But instead for now
It is your solace I seek
At your base I take a seat
My months of respite all complete

I now search your entire dome
For the Hunt trail on which I will roam
To the AT's northernmost
Home sweet home

Red Robin

Chapter 9 A New Sun Rises Every Day
Kevin Conley aka 30 Pack

The summer of 2012 was the first time I truly opened my eyes and began to appreciate the beauty that has surrounded me my entire life. The beauty that lies within the people, mountains, rivers and trees, and through the wildlife that walk, fly, and swim around our majestic planet. I had yet to fully embrace this beauty. Beauty is in everyone and everything if you take your time to see it. It took me 23 years to find my true passions and discover what I value to be some of the most important things in life -- loving yourself, loving the community around you, and showing compassion and kindness to everyone.

They say, "If the trail doesn't change you, you didn't hike the trail." Well, the trail not only changed and sculpted me into a better human being but it saved my life.

Throughout my teenage years and into my early twenties I was a very selfish person. I went from an all-star basketball and baseball player to a punk teenager -- whose primary focus was on smoking pot, chasing girls, and getting drunk with my friends. I ignored the things that truly mattered, like my family and setting myself up for a happy and successful future. I would steal money from my parents to buy pot and sneak out. I was also thrown into juvenile detention centers a few times. For reasons I didn't understand at the time, I decided to find my own trouble. I had more than my fair share of run-ins with the law and struggled with alcohol addiction as a young adult. I was what you would consider a parent's worst nightmare. Looking back, I can truly see that my parents did everything they could to help me. However, change can only happen from within. You need to desire change on your own, for yourself. You must fight for yourself and for your own future. It took me a long time to fight for myself, to stand up for what was right and what I believed in and to become the man I was raised to be. My parents raised me to be a caring, thoughtful, honest, genuine, and kind member of society.

My Grandfather passed away and went to heaven on May 10th, 2010 at the age of 87. During this time my life took a giant downhill turn. My grandfather, whom we called Poppy, was an amazing man. As a boy we would play wiffle ball together for what felt like forever. We would even battle on the monopoly and checkers boards. Sometimes, he would take me down to the pier to go fishing. He always bought me ice cream with chocolate sprinkles, or jimmies as they call them in the Northeast.

I remember the first time I stole something. It was at a pharmacy in Situate, Massachusetts. When we left Poppy noticed I was eating a piece of candy. He asked me where I got it. I don't think I really knew what I did but I told him I took it. I believe it cost about a nickel, so Poppy gave me a nickel and made me go back into the store, alone. He insisted I tell the gentleman at the counter what I did and give him the nickel. This trip to the pharmacy with my Poppy was a lesson on honesty that still influences me.

My grandfather was an all-around wonderful person. He was a man of faith in the Catholic Church. He was also a man of honor as a fighter pilot for the Army Air Corps in World War II in Germany and England. Later, he was stationed in Japan as a US Air Force major in the Korean War. After the war, he went on to be a pilot for 25 years, working for Delta. Aside from work, he also raised three wonderful women, one of whom is my beautiful mother, and one son, Johnny, who passed away too young and too soon.

When Poppy died one of the only things that stuck in my head was the condition in which I last saw him. My Mother had been trying to get me to visit for months. My mom eventually told me that I needed to come soon because it might be the last time I saw him, so, I finally went. When I arrived to visit him, Poppy wasn't really there. That was the last time I saw my Grandfather alive.

When I showed up I brought some chocolate bars. Chocolate was his favorite and it would always put a smile on his face. However, he didn't smile, he didn't talk, he just looked at me with his big blue eyes. We attempted to play a game of checkers but he didn't really move, so my Mother moved the pieces for him. I know he knew who I was, but I never really got to say good bye. I remember leaving that day and beating myself up in my mind, wishing I had come to visit sooner. It was very upsetting and this last memory is burned into my brain -- the way he just looked at me and didn't communicate. He couldn't even hug me back. Why didn't I go visit sooner? Why didn't I spend more time with my Grandfather? Why did I wait so long? Why didn't I love him like he always loved me?

It's amazing how we look back and wish we would have done things differently, or that things could be different in general. However, we cannot change these things. What we can do is recognize what we can do differently in situations like these in the future.

I love you Poppy.

Death is a very difficult process to deal with. Everyone copes with it differently and at different rates. We never truly forget nor should we. It is always hard to let go and move forward. So, for months after his passing I would play old VHS tapes of Christmas, Easter, and fond childhood memories. These videos showed my Grandfather as the cheerful, funny, and alive man that I wanted to remember him as, not the man I saw on my last visit. I still have those chocolate bars I brought on my last visit.

Funny story: Years after my Grandfather passed away I was getting ready to board a flight in Salt Lake City. I have an extreme fear of flying and don't fly unless it's important, VERY important. I joke that I only fly for funerals or women. That's pretty true. Anyway, there is a picture in my mother's room of my Grandpa before I was born, suited up in his Delta Pilot uniform. He looks healthy and young with a bright smile on his face. I can see this picture in my mind when I think of Poppy. So, I was very nervous to fly back to Virginia. I walked up to the counter to ask about the weather conditions and whether it's going to be bumpy. As it turns out, the only person at the counter was the pilot. As he lifts his head up from the desk, the gentleman looks exactly like my Grandfather in the picture that my mom keeps in her room. He had the same sharp wit that my grandfather possessed. I proceeded to tell him that I am afraid of flying. He smiled at me and said, "I am too." It made me laugh and it made me tear up.

That was my Grandfather looking over me. The resemblance was uncanny.

After he passed, I spun out of control. I binge drank for nearly 2 years. I wasn't myself. I would drink in the stark darkness of my basement. I would drink all alone until the sun came up. I would drink until it put me to sleep. I remember the nights that my dad would sit and watch TV with me at three in the morning just to keep me company. In those couple years of drinking I hit what should have been rock bottom more than a few times. I thought about what it would take for me to stop sinking in the bottle of sadness that I couldn't stop pouring. What would allow me to strive to be better? It wasn't the sleepless nights in jail cells or car wrecks. It wasn't the continuous let down of myself or my family. Nor was it the depression and sadness that I rarely spoke of. No, it was a miracle 86 years in the making that really woke me up and made me see the light at the end of this vicious cycle I allowed myself to get trapped in.

I owe a good portion of this rebirth within myself to my Grandmother Shirley and to our friendship and the bond that developed as she was fighting to live. She was the sunshine in my darkness. She was someone that taught me that love is simple. Love is spending positive time with someone and being there for them unconditionally. Before my dream of hiking the Appalachian Trail my Grandmother was in a car accident in rural Maine. She had been fighting shingles for a long time and the doctors had her on 12 different medications, including OxyContin, for over a year. The same week in 2010 that my grandfather passed, my grandmother fell asleep behind the wheel. This led to the decision of my parents to move her in with us. Grandma stayed with us for awhile, but being a very strong and independent woman, we eventually moved her into her own apartment. One day she got pneumonia and nearly died.

At this point, it was impossible for us to give her the care she required. So we moved her into a nursing/rehabilitation place called Heritage Hall. It was only a few miles from my parents' house. I know this was a very hard decision for my father.

You see, I didn't get to spend the time I wanted with my grandfather as he was dying. But now, I had an opportunity to spend time with my grandmother before she died. I wanted to show her what type of man I could be. Almost every day after work I would go down to Heritage Hall to visit Grandma. Most of our family lived in the area, so she had a lot of visitors, my brother and me, my mom and dad, my dad's brother Bob, and Bob's wife.

My grandmother was unable to walk for a time, so I wheeled her around. My father often accompanied us. Slowly but surely we taught her how to walk again. My grandmother was a single mother that raised two extraordinary boys that became great men, one being my father. My grandmother spoke French, had a doctorate in cross-cultural studies, was a teacher and traveled to Japan, the Cape Verdean Islands, and Ireland to write her thesis, which is held by the University of Massachusetts Library. She was such a classy, soft spoken, and beautiful lady. You wouldn't believe how her smile lit up the room like the Fourth of July.

During this time, my Father and my relationship improved significantly. We became best friends. We shared a bond and supported each other through the bitter sweet moments at the end of Grandmother's life. For the first time in my life I needed to be strong for my dad and be a shoulder he could cry on.

"Kindness is the language which the deaf can hear and the blind can see."

- Mark Twain

On one rainy night in February of 2012, I had a cab drop me off about a mile from my house. I don't remember exactly why, but I just wanted to walk and to feel the rain. After I stumbled home, I entered my parents' bedroom and sat on the bench at the end of their bed and started crying. My mother comforted me. I told them that I couldn't live like this anymore. The drinking had become too destructive. I told them I had to do something different.

I was whimpering when i told my parents that I was going to hike the entire Appalachian Trail. I had no idea what to expect or what I was getting myself into. I just knew it was something different and something BIG.

My dad took me to REI the next weekend and bought all my gear. My tent, sleeping bag, head light, stove, sleeping pad, and everything else we thought I needed. I sold my

Jeep Wrangler, and on March 20th my mother drove me down to northern Georgia to take the first steps of a journey that changed me forever.

The Appalachian Trail, Georgia to Maine, March 21st to September 22nd

When it comes down to what happened, what caused such a change, it is something almost impossible to describe. It was many things, but most importantly I learned to live with an open heart. A big part is the people. There is a saying amongst the trails, "The people are the trail." This proved so true. I met so many incredible people who taught me so much along those 2,180 miles of dirt and rock. They taught me in ways big and small. Special shout out to the Wolf pack, the first group of hikers who became lifelong friends and have seen me grow and change. Thank you for your friendships. I will forever be grateful for Bird Shooter's trail magic at Whitley Gap Shelter on my third day, which turned a group of hikers into a family.

The Appalachian Trail was the gateway to some of my first moments of being truly alone. There were nights all alone under the stars. Only my thoughts, the breeze, and my sleeping bag. So many miles, so much time thinking about my dreams. What they are, who I want to be, and what do I want to do with this one life I have.

I definitely had some hard days on the trail. One night, while sleeping in one of the only half-dozen shelters I slept in, I got bit by a black widow on my butt. That was painful and put me off the trail for a couple days. I also got an abscess on my lower back, where your pack rests

above your butt, which ended up getting infected. I got
sent back to the hospital and off trail for a couple more
days. And, to top it all off, I got two stress fractures in my
right foot with 500 miles to go! I took a couple days off,
then powered through the pain. And trust me, it's painful.
But, as the saying goes, "No Rain, No Pain, No Maine!"
Months after the AT, I found out I had Lyme disease,
which is more brutal than one may imagine and
something you should take very seriously.

I summited Mount Katahdin on September 22nd. It was
the first time I cried tears out of joy. I finished with my
dear friends Rampage and Gumbo, who are still very
close and dear friends. Hiking over 2,000 miles, covering
14 states, the views, the history, the people, the dirt, it
transformed me and opened my world up to a life of
adventure.

There is a special bond you make on long trails and on
adventures. Bonds that are created through hiking in
miserable cold winds and rains, through watching sunsets
on the famous Balds of the South, from being away from
technology and making up your own games. Talking for
hours and hours as you climb and descend mountains. In a
way, we were all kids again. It was the beauty of the trail
and the toughness of it that birthed deep friendships and
loyalty. The trail is so kind. The people are so generous,
full of smiles. The trail is more than a community, it is
truly a GIANT family and we always take care of
family. You'd be blown away by the love that you
experience in this world when you have a positive attitude
and live with love. This world loves you back. The trail
provides. I have endless stories of people picking us up in
rain storms and giving us hot showers, food, and a bed for
the night, and asking nothing in return, except maybe a
story or two.

"Nothing great was ever achieved without enthusiasm."

- R.W. Emerson

After the Appalachian Trail I had a better idea of what was coming next. My next adventure would be the Pacific Crest Trail which I had never even heard of until I was on the Appalachian Trail. So, how did I make that happen? About a week after I got home from the AT, I got a job as a machinist and started saving money for the PCT.

Now, let's rewind a tad. When I was half way through on the AT, I went to my parent's home near Harpers Ferry and took a few days off the trail to visit Grandma. I didn't know it at that time but the doctors had told my father she might only have a few days left to live, maybe a week. Somehow, we all rallied Grandma's spirits and strength. She lived longer than expected. My father didn't tell me that news. He knew that if I knew what the doctors said that I would not leave. I would stay with Grandma. My dream of hiking the trail would be over. He saw that I was changing, perhaps even more than I realized. I thank him for hiding that from me at that time, because he would have been right. I would've stayed with Grandma.

When I got home from the AT, I was right back to working and visiting Grandma whenever I could. On my first visit, after climbing Katahdin, my dad and I popped over and brought some flowers. We wheeled her out to the courtyard. It was a sunny and beautiful day. As soon as we went outside, two planes flew over. Their contrails left an X in the sky right above us. I believe you are always right where you need to be. The X, well, it marked the spot.

Grandma was very impressed and proud of my hike. She loved the pictures. Although the dementia was pretty intense at times, I know she was proud. Her biggest fear of me hiking the trail was that a bear would fall out of a tree and land on my tent. I laughed at this but she was serious so I had to tell her a little white lie about how there were no more bears. Grandma never remembered what she had for lunch but she always remembered me. She would make me feel so special when I walked in the room. We became best friends in that nursing home.

My wonderfully inspiring and tough Grandma passed away on January 22nd of 2013. That morning I received a phone call from my Uncle Bob at about three a.m., but I didn't notice it until I woke up for work. Uncle Bob had never called me before, so I knew. As I walked up the stairs my Mother didn't even need to say a word. Grandma was gone.

It was the hardest moment of my life. It hurt deeply. The previous day I was supposed to visit Grandma with Dad after work, but I was tired, so I said I would go tomorrow. There was no tomorrow.

I thought about that and how I was lucky that I got to spend so much time with her. You never know when will be the last time you see someone. Tell people you love them. Leave people happy notes. Call people when you think of them. Try to communicate and work through things to better your relationships. Do special things for people you love, and strangers. Live each day like it could be your last. I try to make a stranger smile every day.

When Grandma passed it was tough. I wanted to go back into old ways and drink. But this time I was stronger. I was able to cope with this in a healthy way, unlike when my grandfather passed away and I drank uncontrollably. Neither of them would want me to go back to that. Now they were both looking down on me. So I didn't. I hunkered down, spent time with my family, focused on work and my dream to get to California and hike the Pacific Crest Trail.

I decided to dedicate my hike to Grandma. I also wanted to do the trail for a greater purpose, so I set a goal of attempting to raise ten thousand dollars for the children's charity, For Love of Children.

On April 25th of 2013, I took my first steps from the Mexican border and started walking north to Canada. The Pacific Crest Trail is 2,650 miles, a backpacker's dream. To be able to hike the AT and the PCT back to back was something special. The PCT starts through southern California's grand but hospitable deserts. Then it goes on to climb through the magnificent Sierra Nevada range and peaks out at 14,505 feet on Mount Whitney, the highest peak in the lower 48. Inching its way up California's endless forest before breaking past Mount Shasta and onto the promised lands of Oregon. It passes by volcano after volcano until crossing the Bridge of the Gods and the Columbia River into Washington, where we chased winter along the spine of the Cascade Mountains into Canada.

The Pacific Crest Trail was the greatest adventure of my life. I had never hiked in California, Oregon, or Washington. The trail was the sweetest of sweet. I fell in love with the trail and I fell in love with the strongest, most beautiful, and captivating woman I have ever met. Her name is Alex.

I want to share with you two journal entries I wrote about the trail.

> As I sit 500 feet above a granite-lined lake, I daydream about moments with my grandmother, a

lady of determination, compassion and love, and a spirit that will always live on in me and through my family. She unfortunately passed away in January of this year. It has been a tough time for me coping with her being gone and I think about her frequently, but also happily!

My Grandma taught me so much when she had so little. I have fond memories of her as I was a child running on rocks on the shores of Cape Cod and the lesson of always finishing your plate off at the dinner table. I have memories of her contagious smile, her blue eyes that orchestrated love and her thoughtfulness towards my brother and me. However, my greatest memories and the most grateful moments we shared were at the very place she left this earth and on to greener pastures. This was at Heritage Hall, a nursing home a few miles from my parent's home. A nursing home can be a dark place but this was a place that lit up my world at a time it was not. I watched my Grandma learn how to walk again. I watched her smile and light up when I would walk in. I saw the simple satisfaction she received from a small bowl of chocolate ice cream. I witnessed her beautiful smile on days where I knew she was suffering. I was blessed with the opportunity of hearing stories from her lifetime, some repeatedly and some out of the blue that were never heard beforehand. I felt so much love when we would hug tightly for a few minutes and when I would feed her. I loved tucking her in and knowing she was warm and cozy before I left.

She taught me to smile, smile through everything. That lesson is a gift she bestowed on me. She taught me a new meaning to friendship and through ours I realized there was more to life than money, cars, houses, and degrees. I learned about love! I left her a note on the summit of Mt. Whitney, the highest peak in the lower 48 and maybe the closest I will ever be to her on this planet. I tell you this because she is one of the reasons I am so alive, happy and free today!!

My Grandmother and Grandfather are still with me through every step. They are alive in the trees, the birds, the coyotes, the clouds, and the rivers.

"Do anything, but let it produce joy."

- Walt Whitman

This next one is a love story I wrote after completing the trail, it's about my best friend and the best hiking partner I could dream of.

Love.

A crazy, beautiful thing.

I remember the first time I saw her, thirty miles into the trail. It was hot, not a cloud in the sky. I was part-way through a climb when she appeared around a bend, breaking on the trailside, her auburn hair striking and her blue eyes dancing in the sunlight. She has that captivating smile -- the way her cheeks rise and her chin moves a bit. We exchanged a brief hello on the mountainside and I kept walking; later wondering why I hadn't stopped to talk with that cute girl.

I began my walk from the Mexican border without expectations, searching for nothing in particular except for that sense of happiness. My only intention was to walk. I set out to hike the PCT alone, as did Alex, but as our friendship and attraction grew we ended up hiking most of it together. She told me early on that she wasn't looking to get into a "relationship". From that moment I worked at it, chasing and charming, hiking my biggest miles to catch up to her and my shortest when she was behind, doing everything I could to get her.

Our first kiss was in the bathroom of a cheap motel, Cajon Pass, mile 342. It was my 26th birthday. A pack of hikers had crammed into a few rooms, hitched 40 miles to town for food for a cook-out, and basked in the luxury of an outdoor pool.

It was nearly 1,500 miles later, just past Crater Lake in Oregon, that we exchanged "I love you" for the first time.

As hiking partners we were a perfect fit. We had long talks; we had silly talks; we passed the hours playing twenty questions, solving riddles, and reinventing the game of Jeopardy. We imagined our own Apollo space missions and I laughed at corny Laffy Taffy jokes while she rolled her eyes. While I scared off what were sure to be bears and mountain lions in the darkness of the forest, she made sure that we always had cocoa as the nights grew cold and the seasons evolved. We developed a system, setting up the tent, building the fire, retrieving water, and through it we learned to care for one another.

Our relationship weathered the monsoon-like rainstorms in Washington and soared as we

ascended Mount Whitney. It was boosted by the
beauty of the trail, the magnificent waterfalls
cascading from the sheer, granite walls of the
Sierras and the sun setting behind the Northern
Cascades. But it was the difficulty of the trail that
most reinforced our relationship. When I broke my
foot in Northern California, she walked behind
me. Her presence was soothing as I winced
through those painful miles.

She comes from Montana, a state from which I
couldn't name a single town before we met. I like
to poke fun at her small-town norms and she
scorns my suburban upbringing. Our differences --
cultural, political, geographic -- enrich our
relationship. Through feet of snow and the intense
desert heat, we encouraged each other to stay
positive through the hardest times and, through
our differences, found a balance that fueled us
northwards.

Every time she walks in and gives me that smile, I'm
taken back to that mountainside. And knowing now what
I didn't then, her kindness, her compassion, her smarts,
I'm captivated.
I still get lost when we kiss.
We didn't start the Pacific Crest Trail together but we
ended together.
Love.
A crazy, beautiful thing

So, as you can imagine the PCT was a blast and it was the
best summer of my life, two years in a row. After hiking
the Appalachian and Pacific Crest Trails I was reborn. I
was full of joy, love, and hope. In 2014, the year after the
PCT, Alex (Outburst) and I went on a 20,000 mile road
trip around the North and Southwest, visiting friends,
family, and so many of our public lands. We started a tee

shirt company for hikers called Backcountry Ninjas. We slung shirts out of her parents' Subaru, which we dubbed "Ole Faithful", on our travels. At the end of 2014, I did a crowd fundraising project called Burgers and Love, which raised a little over $1,000. I hand-delivered 900 cheeseburgers to homeless people on the streets of Seattle. It was truly a beautiful and humbling experience.

At the beginning of 2015 I went to Southern Utah to be a field instructor at a wilderness therapy program called WinGate. It was such an amazing opportunity to help teenagers and young adults. I was able to relate with their struggles and guide and mentor them. It was one of the most rewarding experiences of my life. I believe I made a positive impact. I really cared for all of the young people I met. I loved watching them grow and change in front of my very eyes. The main reason I applied for this job was because when I was a teenager my parents sent me to a wilderness therapy program. Actually, a few times. So, I thought from my experience with it as a youth and with my experiences as an adult and through the dramatic changes I had made, I could really help some of these struggling teens out. And, I did. I accomplished my goal. I am forever grateful for the opportunity.

"Devote yourself to loving others, devote yourself to your community around you, and devote yourself to creating something that gives you purpose and meaning."

- Morrie Schwartz, *Tuesdays with Morrie*

Looking back on the previous five years of my life, I can say I have truly LIVED! I have hiked close to 8,000 miles with a majority of that coming from the Appalachian and Pacific Crest Trails, slept under some of the most open skies for star gazing in the Southwest, where the Milky Way seems to be just a fingertip away. I have been dazzled watching melting sunsets in the Northern Cascades and the High Sierra Nevadas. I've snowshoed to hot springs in Idaho in the cold of winter, I've hiked through rivers, around glaciers, to the tops of mountains, in canyons and caves, on ridges and over saddles. I've

161

seen countless bald and golden eagles soaring above giant sequoias and the firs of the Northwest. I have heard the echoes of owls alongside native petro glyphs in canyons as the water slowly passes through the sandy bottom. I've witnessed endless amounts of love from complete strangers, trail angels, and people from all over. I've felt the energy of the giant redwoods of Northern California. I've crossed from rim to rim to rim of the Grand Canyon, 48 miles in 16 1/2 hours, and I've jumped behind waterfalls and swam in tarns and cirques.

I've cried, sang, dance, kissed, laughed, and meditated at some of the most beautiful and majestic places in the country. I've been to almost every national park in the country and countless monuments, forests and BLM lands.

I've hand-delivered over 1,000 cheeseburgers to homeless people and worked with a ton of wonderful young teens and young adults through wilderness therapy.

I am forever grateful and will always push myself to do more, to BE more, to become a stronger and healthier man, help others as often and as much as I can, and love with all my heart down every path, and through it all remain positive no matter what happens. I'm so in love with life and it's only going to get better.

Dream BIGGER, Laugh HARDER, Love LONGER. And be weird, be you, be awesome!

"The sun shines not on us, but in us."
- John Muir

Happiest of trails,
		Kevin '30 Pack' Conley

~ ~ ~

I was born and raised in northern Virginia close to the Appalachian trail and the Shenandoah Mountains. I didn't discover my passion for the great outdoors until my early twenties. After many ups and downs through life's trials and tribulations, I learned to love the world and people all around me, hug trees, be kind and compassionate to all, and remain positive. Life is what you make it and I want to live a life worth living while also inspiring others.

Chapter 10 Seeking Spirituality
Kerry Smithwick aka Scribbles

"...the soul may learn something of what rest is, as day
after day one opens one's heart to let the sweet influence
of nature's sabbath enter and reign"
God is Enough

Heart beating fast from the adrenaline rush...neon greens
on the trees just budding, blue sky, creek flowing, dappled
sunlight through the pines. Slight breeze on my face – the
breath of God. Slow down. Breathe. You are actually
here.

Garry hiked the first mile up to Springer Mountain and
back down to the parking lot with me.

7 April 2011 at about 11:30, I was 55 years old and given
the opportunity of a lifetime – thru-hiking the
Appalachian Trail.

That was the beginning of a journey that changed my life.

Popular thinking on the Trail is that everyone who hikes is seeking something. Many of the people who start out are 'transitioning'. It can be a transition from work to retirement, from married to single life, from high school to college. Increasingly there are great numbers of people who decided to hike because they have lost a job and can't find a new one. Then there's a group of us who are transitioning but don't realize it.

I took a sabbatical to hike the Trail with every intention of returning to work. It was the right time in my life. My daughter, Danielle, had hiked from Springer to Pearisburg the summer after she graduated from college. Our plan was to thru hike together for our 30th and 55th birthdays. As plans go sometimes, that didn't work out. I decided to go alone. I was working for a family owned business; they graciously agreed to my request and sent me off with blessings. I brought one of the ATC maps of the Trail to the office and they followed my Trail Journal updates to track my progress.

As the days went by I adjusted to living in nature. I've been active all my life – running, walking, biking, paddling, skating, swimming – whatever, I like it! Still, that didn't prepare me for walking up and down mountains all day with 20+ pounds strapped to my back. I was determined to keep a positive attitude. I had signed up to climb mountains and I would do it.

I also adjusted to being alone.

Although I am a loner and enjoy being by myself, I've never been just me, by myself with no responsibilities for anyone other than myself.

I liked it! It was invigorating and fun.

If I wanted to get up early and watch the sunrise – okay. Eat whatever whenever – okay. Stop in the middle of the afternoon just because the view is awesome – go for it!

Of course, I had a very carefully prepared plan to follow – complete with a spread sheet with dates and details. It became useful just for the sheer joy at laughing at how crazy it was to put in that much planning. Okay, I did use it as a guide for when Garry should mail resupply boxes and when we would meet in Harper's Ferry. But, as the saying goes, "If you want to make God laugh, tell Him your plans."

My one objective was to finish. I pictured myself on top of Mount Katahdin. Before I started hiking I was thinking about how I would pose for my summit picture. There was never a doubt in my mind that I would be successful. I would do the miles and finish.

Along the way I encountered all the expected challenges of hiking: rain, cold, ice, no water to drink, too much water because of rain and a hurricane; shortage of food, what felt like endless climbs, steep downhills, a stomach bug, and a spider bite requiring a doctor.

Thankfully, no blisters, foot issues, injuries or serious illness.

The one thing I hadn't planned was my spiritual journey. Not that it can really be planned. Perhaps I should say I didn't plan to read scripture every day or do a devotional. I do pray and I did so when I was hiking. I wasn't actively seeking God, yet I was in nature and acutely aware this beauty and wonder was created by God to give glory to Him and pleasure to us.

Let's consider "seeking spirituality" for a moment. I am a Christian, born and raised in a Catholic family. We attended church regularly when I was young, although less so in my teens. I quit going to church after I was married, and it wasn't until my daughter was born (I was then divorced) that I felt a calling to go back to church. Church was where I found comfort and hope. I married Garry, a good Methodist guy, and began going to church with him. My faith changed over the years and now I cannot imagine a day without actively seeking God's love and guidance.

However, spirituality has many faces. It is different to everyone. It can be organized religion or it can be something that feels much more personal or private: yoga, meditation, quiet reflection, private prayer or long walks. Spirituality involves a sense of connection to something bigger; a search for meaning in your life. It can be an experience or an awareness of your personal growth – something that transforms and/or improves your own vision. Looking within and finding what you are seeking within yourself. Maybe finding a part of yourself you didn't know existed that transforms how to look at life; inspiration that brings joy and love, peace and service.

I found that different spirituality when I was hiking. Yes, I prayed daily. Sometimes more fervently than others – like when I thought I would drop dead before I reached a summit. Or, the first night I slept totally alone – not in a shelter area or where other hikers were tenting. I was okay until I heard some loud rustling and just knew it was a bear. I did look out – it was dark…duh! Nothing to see and no more noise. I laid down and prayed for God to send my guardian angels to surround my tent and keep me safe. Next thing I knew it was morning and I'd slept through the night without incident. I said that prayer every night till I finished the Trail.

As I said, I wasn't actively seeking God. We had a relationship and I was content. Or I thought I was until things that happened in the past started popping into my thoughts. The first was Lee Arrendale State Prison, a women's prison in North Georgia, where I had been serving on Kairos teams. I loved this prison ministry. It gave me an opportunity to serve and be served by a group of women who are often ignored, belittled and have little hope. I found a different type of spirituality when I was with them – one of looking for inspiration and strength from within and from those around me. What I didn't realize was how deeply I felt their injustices and injuries. I hashed this out with God. I was mad for these women and wanted God to fix it.

This became the norm. I'd be bopping along enjoying the Trail: the views, wildflowers, sun, sound of the creeks, birdsong, whatever, and then I'd become aware that in my subconscious I was replaying an old incident. One I thought I'd resolved or let go of. Sometimes it would be things of joy – like Garry and me – that I could give thanks for, but, more often, it was a situation that caused me pain or grief. I would find myself reliving those incidents – crying, sometimes yelling – but coming out of it with a new perspective. A new understanding. And, hindsight always being 20-20, now I truly have left all those incidents behind. (Keep reading, and you'll learn how I know that.)

I look back at my 171 days on the AT and realize how much I grew spiritually. I learned so much about myself. How I was seeking to feel complete and fill holes with physical things. The kindness and goodness of other people played a huge part in my spiritual transformation: finding a cooler full of peanut butter and jelly sandwiches just before the sky opened with a huge rain storm and it would have been impossible to make a sandwich; shortly before Damascus, in the rain, coming to a full shelter and everyone moving over enough to make room for me; meeting Miss Janet, who gave me a ride and remembered my daughter, Mama D!; years of hiking wisdom from Slo Go'in; hours of stories to take my mind off tough terrain by Mark Trail; shared food in the GSMNP from 7 guys who were out for a week; rides provided to and from towns for resupply; the fresh boiled lobster and corn near Bethel and the lady in town who gave me a room with a tub and bubble bath – she said I looked like I could use it.

I summited Kathadin on 24 September. Alone, as I started. It was bittersweet. That was a moment I would have loved to share. The fulfillment of a dream and the celebration of an awesome journey.

Returning to "real life" was shocking. First, to go from walking in the beautiful mountains of Maine to riding a bus, then in a cab and finally a plane within 48 hours of completing the Trail was rude. I sat on the sofa in our den for three days totally overwhelmed – it was sensory overload and way too many choices. It was then that I realized I had drastically changed. Not only did I look different, I was different in my heart and soul. I didn't want all these possessions. I didn't own them, they owned me. I wanted to be free of those ties and I wanted to be back on the AT.

Fast forward 5 years. 6 April 2016, I'm now 60 years old and at 8:30 I started my Pilgrimage on The Camino de Santiago from Saint Jean Pied de Port.

Totally different expectations and objectives. I'd heard of the Camino and knew it was one of the three historic Christian Pilgrimages, the other two being to Rome and Jerusalem. I'd seen the movie, *The Way*. Most influentially, a hiker I met on the AT, Hippy Kippy, has hiked the Camino multiple times with his students from Christopher Newport University, as well as others.

I had not been sedentary for the past 5 years. I walked 110 miles of the Mountains-to-Sea Trail before a leg fracture took me off. I then hiked Glacier National Park and on to part of the Continental Divide Trail before my hiking companions had to quit. I thru hiked The Florida Trail and then the Overseas Heritage Trail, aka The Keys, as well as about 500 miles southbound on the AT.

I wanted to experience a different type of walk, a Pilgrimage. I was attracted to what I thought would be a spiritual – read that as religious – journey.

I have changed a lot since I hiked the AT in 2011. Five years doing sporadic contract jobs, including hand-picking grapes at The Biltmore Estate, and a lot of volunteer work. We built our new home; Garry was the general contractor and I was responsible for everything else. We've really changed our lifestyle – given away 75% of our possessions to live a simpler and happier life. I honestly do not feel like anything owns me. God is at the center of what I do and my aim is to follow His greatest commandment: Love one another as I have loved you.

My preparation for this journey was also different than for wilderness hikes – no tent or associated paraphernalia. Knowing I would be
spending the night in a hostel, albergue, or hotel certainly lightened the pack weight as I did not need to carry food. I did end up carrying more food than I anticipated, however, due to my eating times not matching up with the Spanish eating times and that I get hungry when I hike.

Also, there are several popular guidebooks for the French Way – the most popular of the Camino routes – which break the walk down into between 30 to 35 days with the places to sleep, eat, and visit annotated. That was about as detailed as I needed to be for a walk from town to town.

Because I set out to walk The Way as a Pilgrimage I spent most of my planning time collecting inspirational quotes, reading guide books with details on churches, convents, monasteries, crosses, sacred sites, and history.

The journey to the journey was a journey…. I now live outside of Asheville so a flight to Atlanta with a layover before the flight to Madrid; early morning arrival – find the bus stop for the transport to the train station, layover till the late afternoon train, train to Pamplona, pick up at the station and drive to hostel, spend the night, up early and drive to St Jean Pied de Port – finally get to the Camino office for my official start - having my Pilgrim Passport stamped – hard stop – the passport I have isn't official. Buy a new one, get it stamped, take a picture and I'm off; except the weather is still too risky high in the Pyrenees so the Route de Napoleon is closed. Believe me, the alternative route, which involves quite a bit of road walking, is still challenging -- steep climbs and no food available. I'd failed to heed my own 'note to self', which, by-the-way, was written in my guidebook, of purchasing food before leaving SJPP. Thus my spiritual journey began with me doing a lot of praying to make it up what felt like steep hills.

And the blessings began. I spent the first night in the Roncesvalles Albergue. They've been doing this for years and it's a smooth process getting registered and into your bunk. The Albergue can sleep several hundred people in pods of 4 beds (2 bunks). My 'bunkies' that night were a Lutheran minister on his second Camino, this time with two new knees; a young woman, Katie, from County Kerry, Ireland; and another woman who didn't give her name or location and appeared stressed. Every evening the Albergue, which is attached to the Royal Collegiate Church of Saint Mary, holds a Pilgrims' Mass during which a Pilgrims' Blessing is given. Other than the greeting and a brief part of the blessing, the Mass was in Spanish. St. Mary is a beautiful small stone chapel. In this place, steeped in history, Communion was served by candlelight and it was such a nice ending to the first day. I felt blessed.

All along The Way are stone and iron crosses. Most every town has one and there are crosses in some of the fields or a small altar or niche with a statue of Mary.

In Zabaldika there is the 13th Century Church of San Esteban and a convent at which I was hoping to stay the night...nope, not open till later in the month (which is something the guidebooks do not mention). However, one of the nuns came out and opened the church, turned on the lights and beautiful music, gave me English literature on the church and invited me to go up the old stone winding staircase and ring each of the three bells. What an amazing blessing! I love church bells and am so sorry that in most of the US we don't ring church bells anymore. To

172

be able to do so in this small mountain town, hearing their echo, was so special.

The literature given to me by the nun at San Esteban included an "Our Father for Pilgrims":

> Our Father who is on our Way, may Your breath come to us and watch over us Pilgrims. Your will be done in the heat as it is in the cold, assist us in our weakness as we assist those who falter on the Way. Lead us not into heartbreak, and deliver us from all evil. Amen.

My experience in Pamplona was good but frustrating. The weather was not good. It's a big crowded city and I felt I was wasting time attempting to find the sites I wanted to see. Sometimes I'm searching so hard, over thinking or over planning that I'm missing what is happening right now. My plan was to spend another night and hike out the following morning; instead, I left after lunch and stopped at the Pamplona Cathedral – just to see the inside – Mass had just started and I took that as I sign I was meant to be there. Another amazing blessing! The cathedral is a magnificent 14th Century gothic structure and considered one of the most beautiful in the world. Five priests served the Mass, singing in Spanish, with amazing acoustics. There were three pilgrims attending the Mass and one of the priests gave the Camino Blessing. While I should have been concentrating on the Mass I found myself, instead, seeking inner peace for what I was doing. I realized what I was feeling in Pamplona was a heavy weight – big cities are not what The Way is about, it was definitely time to hike on.

I felt immediate relief once I was through the concrete jungle outside of Pamplona and into the countryside before Cizur Menor.

The blessings were numerous and came in many guises, such as dinner with Kate, a veterinarian from Canberra, Australia. Totally enjoyed talking with her. We had dinner at a local restaurant where a big family was gathered to celebrate the Baptism of a baby girl. Sunrise over the Pyrenees. Church bells ringing on Sunday mornings. Sunlight through stained glass windows just as I was thinking it was too dark to see anything.

In Puente la Reina I had the nicest waiter at Bar Zenon. He had such a kind spirit. Between his broken English and my very broken Spanish, we had a great conversation.

Sharing a communal meal in Albergue Virgin de Guadalupe of lentil stew, homemade bread and yogurt with fig sauce, all served family style with special people. David from California, a professor who works with large marine mammals. Ivette and Nicholas from Texas, two of the dearest people ever. And Tom, from the UK, who had just lost his wife Katrina.

Hearing cuckoos. Seeing storks and their enormous nests.

Walking into Burgos in the pouring rain, I was not looking forward to another big city but I was looking forward to a real hotel room with heat, hot water, and no snoring. I was soaked to the skin and stopped under an awning to check my map with the hopes of finding the street my hotel was on. A sweet older gentleman, who was walking his dog, came up and asked if he could help. I showed him the name of the hotel and address - bless him! - he knew exactly where it was and gave me directions – some in Spanish and some in English – enough that I knew where I was going and walked right to it. God is good!

In that same rainstorm I lost my hat. And, I left my current adapter in the hotel,. When I walked into Castrojeriz two days later there was a trekking shop run by the sweetest older gentleman. He helped me find both items. After I paid, he took both my hands in his and gave me a blessing. We both were misty eyed. I was so happy when I left I went to the café next door for tea and a chocolate croissant – definitely delicious!!

I don't recall where I found this simple poem but I love it!

At the end of the day
Your feet should be dirty
Your hair should be messy
And your eyes should be sparkling

Walking on the old Roman Way, Calzada Romana, I was following in the footsteps of Emperor Augustus. It's amazing to be where literally millions of Pilgrims have walked.

In Mansilla de las Mulas I had dinner with Michael Angelo, an Italian who lives in Birmingham, England, and Karen, a student at Evergreen College in Olympia, WA. I met them outside of a hotel that has a restaurant with a not very good menu – we all decided not to eat there and they invited me back to their hostel, El Jardin, which had a very busy restaurant. It had good food and great wine from the del Duero region. I love how small the world is – I'd never heard of Evergreen until I met Dream Weaver on the AT, that's where he was headed to school in the Fall of 2011 and, now, here was Karen.

Several blessings were received in Leon, including staying in the Pax Hotel, which is connected to the Convent. My room opened onto the Plaza Santa Maria, which has a beautiful fountain with two cherubs symbolizing the two rivers that embrace Leon. The sheer size of the 13th Century cathedral is amazing and the 125 stained glass windows are stunning – the colors are so vivid. Sitting with the medieval Pilgrim statue in Plaza San Marcos is special. He is seated at the foot of the cross barefoot. The statue is situated so the Pilgrim appears to be surveying the St. Mark Monastery. Prior to the 12th Century the monastery was a pilgrim hospital. It became the headquarters of the Knights of the Order of Santiago in the 12th Century.

The small town of Villar de Mazarife has only one hostel, Tio Pepe. I was blessed to get the last bed. It was here that I met Robin, a Canadian who had recently lost her job, and Mette, a German woman. They were my bunk mates. Two gentlemen from Denmark, Martin and Morton, joined us during a cold, wet, blustery afternoon and evening drinking good wine, eating dinner, and having crazy conversations about all the places we've been.

The blessings continued with a personal tour of the Roman ruins in Astorga by the self-appointed Mayor.

The Camino became much more about the people I met. Although, I have to confess that I didn't realize how deeply I had been touched by that aspect of the pilgrimage until I was home. The Camino is a shared experience.

On the walk I noticed that I wasn't actually having deeply spiritual thoughts. There hadn't been any rehashing of old issues and no questioning God about what path I should be on. What was going on?

I knew You were there, God. Your presence was evident everywhere I looked, in the flowers, trees, streams, sky, wildlife, and the kindness of people. I began to wonder, why are You not talking to me?

When I look back at my journal I see I was having interesting conversations with myself – just not the arguments I'd had on previous hikes.

One of the conversation involved the question that has been discussed around many dining tables on The Camino – why are you walking – for Spiritual, Sport or Culture? Personally, I set out to walk as a spiritual pilgrimage. However, I will admit there is some sport to the walk -- obviously, since I was trekking in excess of 16 miles every day. Definitely a lot of culture and history involved which I really enjoyed. In addition to the Camino guidebook I had a Kindle book on the culture and history of the Camino. In the evenings I would read about what I would see the next day and review what I had read the previous night so I could remember what I saw that day and write about it in my journal. I also had my list of inspirational quotes.

The quote I was meditating when I realized God was talking to me, and that I was not listening with 'the correct ears', is from St. Frances de Sales, "Do not wish to be anything but what you are, and try to be that perfectly." This made me laugh because I don't know how to be anything but what I am. A hiker from California, David, told me one of the secrets of the Camino is that it makes people aware of their weaknesses. Now, combining that with the St. Frances quote, I had a real laugh. I had been thinking about my weaknesses, like, I really don't like people walking in front of me. This silly quote came to mind, "If you're not the lead dog, the view never changes.' I crave the open space. To see the world, whether woods or streets, unobstructed, with my path clear. That leads me to keep walking when I know I need a break. Or, not stopping once I'm in the clear. It's stubbornness. I know it is. It's a blessing and a curse. I do aim to be perfect. That is the Type-A personality. Even after five years of retirement I haven't given up my drive to succeed at whatever I do. I'm good at being a Type-A!

While I joke about my Type-A personality and my stubbornness, they have been assets. I'm passionate about the ministry I've been called to, and I feel God's affirmation and blessings in it.

Spirituality is a very personal experience. I have talked with other trekkers about its many faces and forms. How it can be an experience that changes perspective or increases awareness. For me, these two hikes brought totally different spiritual connections. On the AT I expected to be inspired as well as physically and emotionally moved by the splendor of creation. I love the mountains. Seeing them literally awakening to Spring was an experience like no other. Walking with Spring – over and over seeing the trees greening, the wildflowers opening, the fawn, cubs and other baby animals. Crisp Spring mornings. Then the forest going into its Autumnal slumber, preparing for Winter. A quiet solitude could be felt.

I was into solitude and discovering, or rediscovering, myself. I did meet interesting people and, at least one, will be a lifelong friend. The spiritual aspect was more about me reconciling the things I've done in my life with who I want to be. I don't care about leaving a legacy. I do care about honoring God and doing good things in His name – caring for His people. His greatest commandment is, "Love one another. As I have loved you, so you must love one another."

Having never been to northern Spain I had no expectations about being wowed by the terrain. I knew it was rolling hills and hoped the Pyrenees would be beautiful. I knew the "Way" travels from village to

178

village, and I would walk through fields, woods, and streets.

What I discovered was the beauty of northern Spain. How magnificent to see the snowcapped Pyrenees visible to my right almost every day. I love history and was immersed in it every day – it never ceased to amaze me how old buildings are in Europe – stone buildings, thatched roofs, cobbled streets. Miles and miles of stone fences all built from the labor of clearing the fields so crops could be planted. Fantastic wines and great foods. I enjoyed moving through the regions and tasting how the food and wines changed – from heavier, heartier foods and red wines to seafood and white wines. The people were generous and friendly. I have so much admiration for their patience and acceptance of thousands of people walking through their villages, fields, grocery stores, churches and shops – most of whom do not know any Spanish or have a clue about the culture and history. What a mind-expanding trip, starting in Pamplona, learning about all the invasions and struggles Spain has survived. Reading the history and seeing the stone-fortification walls and structures built to defend the cities, it's not hard to picture knights in the medieval towns protecting the Pilgrims and locals.

When it's all said and done – it was about the people I met. I have a special place in my heart for those I got to spend time with and learn about their lives and why they came on the Pilgrimage; everything from "I'm doing it for my mother"; "my husband just graduated from medical school and we are starting a new chapter in our lives"; to "I lost my wife" or "I lost my job" or "I just retired". And God. I met Him in a new light.

The Pilgrims' Blessing given at the Cathedral in Santiago took on a special meaning for me:

Father God we ask Your blessing.
We are pilgrims who have come to venerate the
tomb of Your Apostle Santiago.

As You kept us safe on our Camino way,
May You keep us safe on our journey home.
And, inspired by our experience here,
May we live out the values of the Gospel
As our pilgrimage through life continues.

We ask Saint James to intercede for us as we
Ask this in the name of Jesus Christ,
Your Son and our Redeemer.
Amen.

So, I didn't end up back on the AT, and I didn't go back
to work full-time. In February 2012 I started hiking the
Mountains-to-Sea Trail from Jockey's Ridge State Park
on the Outer Banks to Clingman's Dome in the Great
Smoky Mountains National Park. Unfortunately, I
sustained a fracture in my left tibia and had to get off after
about 110 miles. That didn't stop me from hiking – I
joined my soon to be new hiking buddy and life-long
friend, Funnybone, as he began his AT thru hike from
Springer to Neel Gap. I hiked that same stretch twice that
Spring – the second time with Mark Trail. Then, in July
2012, I took a short trip to Montana – the actual trip was
long on the train – the hike ended up being shorter than
planned. I did get to hike the Glacier National Park loop
and start on the Continental Divide Trail.

While I was hiking in Montana, Garry moved us to North
Carolina. Our house in North Georgia sold, so our home
is now a cabin in NC. Starting in January 2013 I hiked
The Florida Trail. 2014 I started the Benton MacKaye
Trail but had to get off due to a death in the family. I
decided to start SoBo section hiking the AT in 2015 and
have completed about 500 miles. I have also hiked the

Overseas Heritage Trail and the Florida Keys from Key West to Key Largo. This past February I completed the 71 miles from Key Largo to the Oasis Visitors Center. I convinced Garry to shuttle me so I could complete that section in 3 days – whew! Lots of road walking, alligators and crocodiles. The prize was a long weekend in Key West to celebrate. All that's left now on the eastern coast is to connect The Florida Trail to the Appalachian Trail which involves hiking all of Alabama up to Springer Mountain -- it's on my list along with hundreds of other trails. Once the bug bites it's hard to get rid of that itch.

Hike on!
Scribbles

~ ~ ~

I'm 60, a frequent hiker, and enjoy all kinds of outdoor activities including running, biking, paddling, swimming——whatever. I thru hiked the Appalachian Trail in 2011 and have hiked in Montana, Florida, North Carolina, and 500 miles SoBo on the AT since then. My most recent hike has been on the Camino de Santiago in Spain. This was more like a pilgrimage for me continuing my quest for spirituality that began in earnest on the Appalachian Trail.

Chapter 11 The Bond of Sisters
Renee Neufeld aka Kleenex

I wanted to hike the Appalachian Trail, but I didn't really want to do it alone. This is where Alison, my sister, comes in.

Hiking partner

I tried every way I knew to convince Alison not to come. I sent all of the "worst case scenario" hiking stories: only 12% succeed in hiking the whole thing, it takes 5-7 months to thru-hike, rain is common, the cost is between $3000-5000, expect to be discouraged a lot and on and on. It wasn't that I didn't want my sister to come with me, but this was my dream and not hers. I wanted to make sure she knew what she was getting into. Not only would there be lots of trials just by being on the trail, but she would have to put up with me for 6 months. This wouldn't be the same as living in the same house for 15 years; we would be with each other 95 percent of the time. The other 5 percent made up of bathroom time, getting water or other miscellaneous times apart. I love my sister and we don't really fight about how to hang the wash on the line anymore, probably because we don't live in the same place. The only time we really get on each other's nerves is when we are in a new city and trying to figure out driving directions. Being on the trail would bring a whole new meaning to sisterly bonding.

I was excited to spend this time with Alison. Now that we're all grown up I really enjoy sharing adventures with her. She is my best friend. We would be able to share stories and inside jokes for years to come. We would be able to live and breathe the same adventure. It was going to be awesome.

Of course it's always good to talk with your hiking partner before you begin as to what would happen if one of you had to leave the trail. For us it was simple: If I had to quit, Alison would quit; if Alison quit I would continue hiking. This was my dream after all. Alison was excited to share it with me, but it wasn't enough for her to go on if I wasn't there.

Preparation

We started preparing for the hike in earnest in December. I am a planner, so I made sure we had everything down to a science. If I don't have at least 10 lists going, I feel a bit lost. I guess I'm afraid that I will forget everything. That's why I sleep with a pen and paper by my bed. If I can't sleep because I am worrying about something, I write it down and it frees me to fall asleep.

I have always enjoyed the homemade beef jerky that I took up to the Boundary Waters of northern Minnesota, so Alison and I spent a day at a neighbor's house and dried 60 pounds of beef. All through December and January we coerced our parents to help us taste test the recipes we were going to use on the trail. They were very gracious, but were also very happy when we picked the winners. After we figured out our meals we proceeded to buy fruits, vegetables, and meats to dehydrate. Mom was instrumental in keeping the dehydrator running, as well as helping us measure and put together meals and care packages. Amazon seemed the place to find the weird ingredients that we needed, such as dried coconut milk and dehydrated sour cream. Mom was our trail angel in South Dakota sending our care packages up the trail to different towns. We decided to send mail drops so that we wouldn't have to spend much money on the trail. And, we wouldn't have to subsist on ramen and instant mashed potatoes. We packaged up 90 breakfasts, 99 lunches, and 90 suppers. The chest freezer was full!

We poured over The Thru-Hiker's Companion and other guidebooks. Plan A was scrapped after only mapping out 4 days, and Plan B lasted until we realized we overshot our October 15th deadline by 11 days. Oops! Plan C was declared the winner but we fully intended to have a plan D while we were on the trail. We had every day mapped out – how far to go, where to camp, whether we would be in and out of town or stay the night, etc. We did follow plan C pretty closely early on in our hike.

In December 2011 we did our first test of our tent. Here's an excerpt from our Blog:

Because of the balmy December weather in South Dakota we decide to take advantage of the non-snowy ground and take our tent out on her maiden voyage.

Objectives – see if the tent leaked, see if our sleeping bag ratings were accurate, and see if we fit in the tent.

Weather conditions: low of 32 degrees, rainy (perfect for testing)

In the early afternoon of December 29 we attempted to set up our tent. It took an hour. Oops. Hopefully it won't take this long on the trail. After an ah-ha moment we got the right poles into the right holes and we took a picture of our achievement. We set the tent up so that it could be seen from the road. We thought it would be a good conversation starter over coffee at the Turkey Ridge Store.

At 10:30pm we grabbed our gear and trudged into the rain to begin our experiment. Once inside we marveled at the fact it was still dry and that we had in fact set the tent up correctly. After some tricky maneuvering we got settled and read for a while before turning out our flashlights.

Fast forward to 2 am. Our dog Sunny was whimpering loudly and licking the rain fly on the tent. The rain had stopped, but the temperatures continued to fall. At this point Alison has not slept at all, and Renee has tried to curl up into a tight ball to stay warm. After a short conversation debating whether to stick it out or retire inside we decided sleep was more important. We fully realize this option will not be available while on the trail, but our test proved 2 things: Renee needs a warmer sleeping bag and our tent is actually waterproof and has enough space.

The really sad part of this story is we left the tent outside too long the next day and Sunny decided Alison's sleeping pad, book and our tent fly looked appetizing. So there were bits and pieces scattered across the yard when we went to take the tent down. The dog was tied up for the rest of the day and we are now looking for replacements.

Moral of the story: don't leave anything outside that you may need later.

We ended up getting our rain fly repaired for about $100. A lot cheaper than a new tent!

On our way!

We drove down to Georgia as a family, lots of togetherness and bonding. We met with several groups of friends on the way and in Georgia. The hills and trees were much different than what we were used to in South Dakota. It was all so beautiful. The night before we headed out on the trail I asked Alison to cut my hair. I have never had trouble with novices doing the cutting, but she was a special case. To be fair, she did warn me. Thank goodness mom came to the rescue. It was shorter than anticipated, but hey, I'm not going to see it, I will be hiking!

Our parents sent us off on the approach trail at Amicalola Falls the next morning after some discussion about directions. Before we headed onto the trail Alison and I weighed our packs, 46 and 49 pounds respectively. Yikes! I was a little nervous about that. We had stuffed all the food we didn't send in our mail drops into our packs. We even had Cheese Buttons from the previous weekend's Schmeckfest (a German tasting festival in Freeman, SD) festivities. We decided we would need to eat our way up the mountain so our packs would get lighter. Unfortunately, we weren't very hungry. We registered as hikers number 715 and 716. It was a sad goodbye at the arch to the start of the trail. Dad had tears in his eyes and we had to turn quickly and start hiking lest we would start crying as well. The stairs at the Amicalola falls approach trail took our minds off of our parents rather quickly. We thought we could handle the 175 steps, but soon realized there was another sign with 425 steps. WHAT?!? We stopped at every platform, drinking in the beauty, and tried to remember why we signed up to do this for another 6 months. We stayed at Black Gap Shelter, 7.3 miles of mostly uphill hiking on the approach trail. We hadn't even made it to the "real" trail yet. But, we were on schedule. We didn't have to use plan D yet. ☺ We made a supper of cheese buttons and spiced it up with power gel. I don't suggest doing that. We had to choke it down.

In the campsite we started the tasks that would soon become part of our daily ritual. Purifying water, setting up the tent, hanging bear bags, brushing teeth, and stretching took some time. We decided to tent most of the time, primarily for privacy, but also because why carry a tent and then not use it? Every evening before we went to sleep, we would think over the day and write what good things and the not-so-good things that happened. It was a nice closure to the day. Entries for the first day:

> Highlights: the falls, getting to the shelter; lowlights: hiking uphill before lunch, the stairs and endless uphill. The first day was definitely difficult, but it was still new and exciting. We woke up the first morning groaning due to sore muscles and realized this was just the first morning of many to come.

Easter

As we hiked the next few days we got into a rhythm. It took us about 1½ hours to get ready in the morning from wake-up to starting to hike. I usually made breakfast while Alison taped every single toe with duct tape to ward of blisters and protect the ones that were already there. We would pack up our gear and then be off. About 2 hours into the day our energy stores would become depleted and we would stop for a snack. A couple hours later would be lunch time and another break a couple hours after that. Sometimes we would even have elevensies mixed in there.

Around Jerrad Gap there was a bear canister rule and we didn't have one. So we camped about three miles away because bears can't travel that far, right? We shared our lodging choice with many other hikers with the same idea. A guy we called "loud army guy" helped us finish our Alpine Spaghetti. It was too much for us. We had not gotten our hiker hunger yet. We woke Easter morning to a section hiker with bunny ears on handing out candy. We don't know where she came from or where she went, but the extra boost of chocolate was welcome. Easter Sunday was also the day we decided my hiker name would be Kleenex for all the snot rockets I did. I thought it was better than snot. Alison has been named 3 step, because she takes 3 steps when she could be taking 2, especially going downhill.

A few things we learned in the first few days of hiking:

- We don't like reconstituted dried eggs. We had to tell our mom to take them out of every one of our mail drops. We couldn't even choke them down.
- We soaked our dehydrated veggies while hiking during the day and then it took a lot less time to cook in camp
- Fill up water bottles the evening before to save time in the morning
- When you stay with people more than one night in a row it feels like you're coming home to family when you arrive at a shelter.
- The little things, like blowing your nose with a real Kleenex, are the most exciting
- We haven't seen very many girls on the trail and not two hiking by themselves.

Three days in we were just a couple miles from the Walasi-yi, when we had our first breakdown. Mentally, that is. This was hard. I understand why some people don't even make it 3-4 days before giving up their thru-hike. It is really hard to prepare for something until you've done it or something like it. I was the cheerleader, and convinced Alison to keep on going. We got a good meal and sent some of our items home. We got our first shower of the trail that didn't come from the sky. We charged our electronics, called our worried mother, left our unnecessary items (candy canes) in the hiker box and slept for the first time in a hostel. The decorations were a cross between Halloween and Hawaiian. The resident cat walked all over me most of the night, so it wasn't as restful as it could have been. And we fell asleep to the cacophony of snoring.

A Quote from my journal: *So far **every** day has been hard in some way. I can't wait til it gets easier.* This was written April 13th. My present self would tell my past self, that it doesn't get any easier, something is always hard whether it is aching feet, emotional turmoil or the mountains in front of you. Yet, maybe it was good that I thought it might get easier, otherwise it would be really hard to move forward. Mental fortitude is so important to keep going. Biting off the trail into manageable chunks is paramount. Thankfully, Alison and I never hit the mental breakdowns at the same time. As a marathon runner I have had lots of experience with mental fortitude. My body often says that I want to be done, but I keep going. I think that experience helped me tremendously during the AT. I was invited to a closing circle of some boy scouts one evening in Massachusetts. *One leader talked about happiness and that happiness is the most important thing in life, but don't get in the way of other people's happiness. That isn't my way of thinking. If my goal was to be happy all the time I wouldn't be hiking the trail anymore.*

Al and Ashley

When hiking 2,180+ miles, walking 1.2 miles to a shelter that is off the trail holds no appeal. So, therefore, we decided to do our first stealth campsite at Hogpen Gap. We were content to be by ourselves, taking our time getting water and doing laundry. As we made supper we heard some people coming down the trail. These guys decided to stay with us at the campsite. Partially because we were 2 girls camping alone kind of near a road. It was nice to know that people were looking out for us. We knew we would like Al and Ashley as soon as they started talking about Veggie Tales. It was so amazing how many Christians we met along the trail. Those are the people we connected with most. We met up with Al and Ashley several times over the next week or so at the shelters. We talked, laughed, and played Yahtzee. It was so awesome meeting people that were fun to hang out with.

Town Stay, North Carolina, and Faith

8 days into our trip was our first town stay. I don't know what Hiawassee is like to normal people, but to us it was paradise! There was a grocery store, multiple restaurants and BEDS! Our accommodations at the Budget Inn felt like a deluxe hotel. The large room looked like a haphazard garage sale as we air dried our clothes and camping gear after we decided not to spring for the extra two dollars for the clothes dryer. We each had a KING sized bed to sleep in! It was so nice to have more room after living in a tent and sleeping bag for a week. The shower was amazing as well. I did feel a bit bad for the cleaning crew that works at this hotel. We tried not to leave it too much of a mess. We made appearances at several of the restaurants – Dairy Queen and Daniel's Steakhouse to fulfill what was becoming the "hiker hunger".

We also overnighted at the Blueberry Patch Hostel. It proved to be a healing balm as the owners are Christian. They openly shared about their faith and it was comforting for us. I found that whenever someone prayed for me while I was on the trail I couldn't help but cry. It was so humbling. We also tried our hand at hitching a ride. We had mixed results: a construction worker who drove so fast that a fellow hiker almost flew out of the bed of his truck and a Gideon who shared his faith with us. On the return to the trail we got waylaid by a Clint Eastwood filming crew. Maybe we'll be famous! Our re-supply was mostly from the care package that we received from Mom. It was nice to know that was coming so we didn't have to spend so much time shopping.

The next day on the trail we entered into North Carolina! One state down and 13 to go! Our campsite at Bly Gap was a nice one with good people around. As mentioned earlier we did not see a lot of women on the trail, but we found lots of brothers and fathers. It seems like all the guys are watching out for us and making sure we're ok. It's a nice feeling. Alison's blisters seem to be holding her back; she even wore crocs for part of the day.

Loud Army Guy had been hiking with us off and on. His trail name is Reaper. Alison and I read Galatians 6 about being a good example to bring others to the faith. We believe we are meant to be witnesses on the trail and everywhere and that this young man has been brought into our lives to see our witness. He and I shared Bible verses and prayed on Albert Mountain while watching the sunset one evening. On another occasion we held our own worship service, complete with singing and a sermon. I wonder who else heard us.

Some highlights from this section:

- One hiker snoring so badly and talking in his sleep so no one got any shut eye in the shelter
- Making pudding in the stream
- Hail – twice
- Hiking in the rain and Alison sitting in a puddle
- Nearly sleeping on top of people in a shelter and Reaper sleeping in a hole

Sylva North Carolina

Sylva will always hold a dear place in my heart. The Appalachian Trail doesn't even go through Sylva, but through someone I knew and Facebook we connected with the Gibsons. They were the most gracious hosts for two nights (and beyond for Alison). Reaper stayed with us at the Gibsons' and zeroed there. One of the most exciting things about going to someone's house is getting a home cooked meal. Our meals were pretty good out on the trail, but we didn't pack in pork chops and baked potatoes! It was the best meal we had eaten in a long time!

On our way to the NOC to get picked up by the Gibsons, we had our second emotional breakdown. These always seemed to happen with two miles left before a place to stay off the trail. Alison told me that she doesn't think she likes hiking and feels bad that I have to be Miss Sunshine all the time for her. I told her that she makes the trail fun and it would be hard to go on without her. We eventually made it down, but we knew we would have to talk more about this.

At the Gibsons' I was able to play guitar, talk about South Dakota, and laugh! We met the Woodwards and they blessed us by offering to bring out our care package to the trail when we went in to Gatlinburg. The more people we met, the more we were humbled by the things they did for us! Their generosity blew us away! Someone offered to take us in to town where we could treat ourselves to McDonalds on a cold wet day. Just when we needed it the most, there would be some trail magic! As a bonus, it was Alison's Birthday. What a great birthday present!

A few highlights:

- We saw our first couple of bears in the Smokies
- Laughing through the hard times so we didn't cry

The decision

April 25th entry – *'I am not sure Alison will go much further. She is really struggling.'*

On the 26th we set out from Siler's Bald shelter and it started raining heavily. Then it started hailing. Quarter sized hail was pounding our bodies and faces until we couldn't even hold a conversation for the noise. Reaper ran after us and convinced us to go back to the shelter and get dry and warm. It took awhile, but we did get warm and I convinced Reaper and Alison to press on toward Clingman's Dome – the highest point on the trail. They were not happy with this choice, but there was a section hiker with a truck there that could take us in to Gatlinburg. So on we went.

Along the way we met a couple of women hiking and one was having trouble breathing. We took out some things from her pack and hiked them up the mountain for her, so she could take her time. But then, we got a message that she was still having trouble and was asking if Reaper could come and get her whole pack. So, I took Reaper's pack on my front side and continued on while he went back to carry hers. Other than it was hard to see where I was going, it wasn't too bad to carry two packs. I told Alison to wait at a fork in the trail so people would know which way to go while I trekked down to the parking lot. I hiked back to the trail and then sent Alison down to find Reaper and the two other hikers. We all made it to the pick-up and crammed in for the drive to Gatlinburg.

Gatlinburg was a pivotal place in our journey. This is where Alison decided to get off the trail. It wasn't without tears and apologies, but it really was for the best. Her emotional and physical wellbeing needed her to get off the trail. I was glad that we were able to be honest with each other and figure out what was best for both of us.

When Alison left the trail it was really hard. She felt like she was letting me down and the realization that she was not cut out for the task at hand was difficult. Alison liked everything about the trail except the hiking. ☺ She enjoyed being out in nature, the people, the solitude, and the sister time. It felt better to me to know that she had a place to stay with people I knew once she got off the trail. She would stay with Rose, one of the women who we helped up the mountain. This turned out to be an opportunity to share about her experience with Mercy Ships, an organization that Rose had always wanted to work for. In God's perfect timing, as the result of their meeting each other, Rose ended up serving with Mercy Ships for two years. Even in the midst of our making that tough decision, God used Alison as an instrument to give her the push she needed.

Because I had more camping and hiking experience there was pressure on me to make it work for Alison. I don't think I realized how much I worried about her until she left. It seemed like a weight had lifted, and I could hike my own hike. I love my sister and I'm so glad that she came with me. She was not a burden, the way I felt responsible was the burden. Reaper pushed Alison and made her do more than I probably would have and it drove her to tears on several occasions.

2 weeks

Alison decided to stay around for two weeks in order to slack pack Reaper and I and meet us in towns along the way. It was so wonderful to have a car in Hot Springs! While in the 'Springs it was fun to relate the things that had happened on the trail since she had gotten off. Reaper and I took turns reading books in the Dolly Parton Imagination Library at the diner in Hot Springs. It made for a good laugh.

One of my favorite nights on the trail came at Cosby Knob shelter where we met Stump finder, Grizz, and Gypsy. They invited Reaper and me to a Bible Study. I about jumped out of the hammock I was sitting in. We sang hymns and the shelter was quiet as we worshiped. It was such a surreal feeling, and so life giving. We met up with them several more times and they began to feel like family.

About a week after Alison left the trail Reaper left as well because he was not feeling well. We think it may have been Rocky Mountain spotted fever. It was hard to be hiking alone, yet also freeing. Once Alison and Reaper left plan C for my hiking schedule went by the wayside and it became plan HYOH (hike your own hike). The first 3 days by myself I did 18, 19, and 17 miles. At one point I was five days ahead of schedule. It was nice to go at my own pace. I had lots of time to think, which was cool and scary at the same time. The next week or so Alison met me nearly every night and I only needed to carry one day of food at a time.

Time goes on

May 22nd Alison left for home. I cried. She cried. I walked through tears for the first three miles and then I met another hiker who talked with me and helped me not feel so lonely. I met lots of other people as I hiked mostly by myself for the next 3 weeks or so. Some I saw only once, but others I would see repeatedly. As my mileage picked up I caught up to some people I hadn't seen since the beginning. That was fun. I did find that neros or zeros weren't as fun because I didn't have a close group to hang out with. Alison was planning on meeting me in Waynesboro, Virginia in less than a month. That gave me a goal and I stuck with it.

Family

All along the trail I felt like I had family or close friends with me. Conversations are easier when you're going through a physically challenging situation. Everyone looked out for each other. I was never really afraid. People in towns, when they heard I was walking the AT, would offer food or lodging. I felt like everyone was taking care of me. During Trail Days in Damascus I almost felt pampered! I could do laundry, take a shower, and eat for minimal money. Some volunteers offered to wash people's feet. Now, hiker's feet are not the nicest looking lot, but these ladies gave us a blessing for our nasty feet that would carry us to Katahdin. It was really cool.

The trail magic was amazing too. Imagine coming around a bend in the trail and smelling someone grilling hamburgers. Then you notice the tables with muffins, fruit, corn on the cob, salads, and desserts. You felt like you died and went to heaven! Trail magic is awesome because those are calories you didn't have to carry while hiking up a mountain, and you can leave your trash!!

A few people came into my life for a short time, even as little as four days, but they became family:

Sly Fox and Tex – met them before Damascus and ended up staying with them

Wiffle Chicken – hiked together off and on in VA and beyond

Trekking Pole – did a really fast hike together and talked the whole way

Flapjack and Bison – met Flapjack in the beginning, was with the two of them for awhile in VA, and met Flapjack again in New England

Poncho – helped me get through some days in VA

Missouri Mule – He and I were at the same shelter a few nights in a row all by ourselves. Later I found out he broke his leg in Maine and was unable to finish his thru that year.

Nooga and Bandana – When I came back from Waynesboro I hiked with these two pretty much the whole way through Shenandoah National Park. I had met them before, but because I had been off the trail for three days everyone I knew was farther along. It was so nice to have familiar faces that first night back out. We stopped at waysides, passed the 1,000-mile mark, and endured a windstorm in Harper's Ferry together. I know these guys would have had my back if anything happened to me. It was so nice to have them on the trail. I bumped into Nooga again in Maine and then at the end of the trail.

Bad Penny- I met Bad Penny (from Germany) within the first week of hiking, and then didn't see him for a while. He would just show up every now and again. I probably saw him once a month or so, and I saw him the day he summited. He'd become an old familiar friend.

Comma-Kaze and Blue Sky – I met Comma in June and one of the first things we did was eat at the Home Place in Catawba VA. It was a Sunday afternoon and all these people were dressed up for church and smelling good. Then we come in --stinky, dirty hikers. It must have been quite the sight! I hiked with both Comma and Blue Sky off and on in VA and beyond. It was nice to split a hotel room four ways with them and Castaway. Comma got off the trail a little after Harper's Ferry because she was only planning on doing half of it. I continued to see Blue Sky every once in a while and even in Maine.

Castaway

I met Castaway the night before Pearisburg, Virginia. We were doing about the same mileage, so we decided to hike together. He said later that I came at the right time for him, and I think he came along at the right time for me as well. God brought us into each other's lives for a reason. Castaway and I actually ended up hiking together off and on until a few days before Katahdin when he took a few days off so his foot could heal. The day we left him at the hostel was one of the worst days of my trail life. It was even worse than saying goodbye to my sister. Castaway and I had been together for over 1,000 miles. I wanted him to make it. He did summit several days after I did. It was a tearful reunion! Castaway watched out for me, wouldn't let me hitch hike by myself, and was a person to discuss all sorts of things with. I am so glad that we met and became friends.

One of my favorite memories with Castaway was in Glasgow, Virginia. I arrived a bit before Castaway and picked up my mail drop. I set up in the shelter in town, which was a cement floor with tarp walls. Castaway arrived and opted for a tent spot nearby. Here is a post from my blog:

Last night (June 8th) I was trying to fall asleep on the cement floor in the Glasgow VA shelter, when I started hearing a rustling sound. I figured one of the hikers had come in and was rummaging through their belongings to find something. However, when I looked around I saw a skunk that had gotten into some hamburger buns and other foods left in the hiker box.

What to do!?! Should I try to sneak out of the shelter? But where would I go? How would I know if the skunk had left? I was the only one sleeping in there, so I didn't have any company. So, I decided to stay perfectly still and watch to see what the skunk would do. Maybe it wouldn't detect my presence, or at least not be threatened by it. As I watched it tear into several packages and then become disinterested in that. It started walking towards me! AHH! Breathing was becoming more difficult. It stopped about a foot in front of me and was sniffing the ground, then it turned and it's tail was facing my direction. Great! If I get sprayed no one will stay in the shelters with me. As I was figuring out how long it would take for the smell to go away the skunk casually walked out, as if it had no cares in the world. Once I started breathing again I was able to calm down enough to go back to sleep. I never did get sprayed, but I am eager to get back out into the woods where people don't leave as much food laying around.

At least the skunk didn't try to cuddle with me!

Castaway and I also went to a carnival in Glasgow. We rode the Ferris Wheel and ate shaved ice. It was nice to feel like a real person and do things normal people do. We played Yahtzee off and on during the trip. It was a nice way to relax before bed.

Waynesboro

I was able to take three zero days. Alison drove over from South Dakota to meet me. It was so nice to process the trail with Alison and to be able to rest. I did a gear swap out and Alison took the extras. I traded in my heavy hiking boots for a pair of low hikers, a scissors for my multi tool, and a sleeping bag liner for my 0 degree bag. The space it saved was amazing. She became my bounce box. I introduced Alison to the people I had been hiking with. Alison and I can be pretty strange together and it was nice to be totally relaxed and not have to worry about impressions. We already know each other's quirks. The first day back on the trail I was able to hike with Alison and a friend who was attending Eastern Mennonite University. The miles were slower, but I wouldn't trade being with them.

Meeting up

Alison took a travel nursing job in Vermont shortly after meeting me in Waynesboro. Since she works three 12-hour shifts a week and she was within driving distance of the trail, she was able to meet me several times. I would plan my zero days in a town when she would have multiple days off. This meant I adapted my hiking schedule to her schedule. Some longer days were required, but it generally worked out well.

We met at High Point, New York and stayed at camp Deerpark together. In Great Barrington, Massachusetts we watched the Olympics together. From Rutland, Vermont she slackpacked X, N, Castaway, and me over Mt. Moosalake, met us in Gorham, and at the end of the trail.

One interesting meet-up:

While Alison was in Rutland we decided she would slack pack me about 10 miles the next day. I would do 5 more with my pack. I arrived at the road crossing and I waited...and waited. The problem was that she couldn't get to me because the road had washed out. There was little cell service for her or me and so there was about 2 hours' worth of panic about how to get in touch with each other. I was able to get a text out, but she couldn't get it. I assumed she was having trouble and told her I would hike another 9 miles to a bigger road. I hoped she would get it and not carry my pack to the meeting spot. In the meantime she had figured out she couldn't get to the trail and used someone's land line to call me. I happened to keep my phone on and was climbing a hill so I had better reception. I was so glad we had made contact and I didn't realize how worried I had been until we made plans. So, she picked me up at a road 19 miles total and we got a room at Quality Inn. It was really nice and I got a shower and we had an amazing breakfast in the morning. It all worked out well, but could have been much worse. I was with Castaway and Bad Penny and they would have taken care of me, but I didn't have anything I would have needed for night if I had to stay in the woods. I did have enough snacks to make it through and treatment for my water as well. It threatened to rain and I was thinking....of course rain would be perfect right now. It's always raining in movies when something bad happens!

And all this makes me wonder...is this what it was like before cell phones???

The Troverts

At the end of July I first met the Troverts. Blog entry: *As I was coming into Salisbury CT I met the Troverts (X and N). I assumed Trovert was their last name, but as I heard them introduced to someone else I realized it is eXtrovert and iNtrovert. Ahh! I think those are the most clever names I have heard on the trail. I had long read their entries in the shelter registers and so it was fun to meet them finally.*

201

*I tell you what, this couple has been a blessing to me.
They are witty and fun and great to hike with. When we
are hiking together up a hill we hardly realize it because
we are talking the whole time. They definitely have been a
ray of sunshine in my life. I would love to hike with them
more, and I am sure that I will see them again. Our
schedules will be similar to the end.*

Near the end of August, the Troverts, Castaway, and I
decided to hike the rest of the trail together. New
Hampshire and Maine are two of the most beautiful, but
also most difficult states, and I was glad to have company
through those states.

Some of my favorite memories were:

- Franconia Ridge – so beautiful and X and I found
  a rock like a recliner and rested there.
- The dungeon at Lakes of the Clouds Hut – *Lakes
  of the Clouds Hut is 1.5 miles before Mount
  Washington. This is the largest hut with capacity
  for nearly 100 guests. Because it is close to
  Washington a lot of thru-hikers will stop in here to
  check the weather and possibly do a work for stay
  there. There were 9 of us there as weather was
  starting to roll in. 5 of us got there at 3:45pm and
  when we inquired about work for stay they said no
  because we were too early and they knew there
  would be more that would show up. They couldn't
  in good conscience tell us to go to the next hut (7
  miles away) because weather was coming in on
  Mt. Washington. But, they did tell us we could
  have the dungeon which sleeps 6 for $10 plus $5
  for supper and $5 for breakfast leftovers. We knew
  we couldn't go any further, so we decided that was*

*our best deal. We got to the dungeon and it
smelled like oil and was pretty cramped. It is used
as an emergency shelter in the winter.*

- Susan - a friend of X and N from Mississippi -
  coming to visit us and slackpacking us while in
  New Hampshire and Maine. So blessed!
- The first views of Katahdin. We all got really
  emotional. White Cap Mountain and Rainbow
  Ledges.
- All the talks and stories. Listening to N tell about
  civil war history. Eating lunch with beautiful
  views and packing in muffins for Castaway's
  birthday.
- Camping at a boat launch with a beautiful view of
  a pond. Mr. Childs was so generous with multiple
  beverage choices and snacks. Amazing.

Katahdin

Alison's travel nursing assignment ended the week before
I summited Katahdin. It worked out perfectly for her to
pick up a friend of mine and come see us at the end. *We
had the opportunity to camp with some friends at KSC
(Katahdin Stream Campground) and we decided to
change the reservation to Alison's name, but we couldn't
get ahold of her. We left her a couple of messages as we
held up the phone while standing on a gravel pit near
Abol Bridge. Park Ranger Jonathon helped us figure out
that there was cell signal up there. He was talking to
everyone as they entered Baxter to make sure there were
enough spaces for everyone to camp. September 25th was
a really popular date to summit because the weather was
going to be nice.*

*We finally gave up trying to call her and hoped everything
would work out. X and I started hiking and N went back*

to the store to charge his phone. A few minutes into our hike N came running up and said Alison had shown up at the store. Wow! So, she didn't get our messages, but she came anyway. We told her the new plans and she slack packed us for the last 10 miles into KSG.

The last 10 miles were very pleasant and we took our time, enjoying the streams and sunshine and the last real day on the trail. It was very nice to have smooth trail for most of the day. It is quite surreal. It was nice to have everything figured out with Alison as well.

We got to the campground around 4pm and Alison and Beth were waiting at a picnic table. It is very nice to have a picnic table again after many miles without them at shelters. We signed in at the ranger station and I was number 537 of the northbound thru-hikers this year. I guess last year there were 585 total for the year. They will definitely surpass that. I wonder how many south bounders will finish too.

Bed time was around 7:30pm as per usual with expectation that we wouldn't sleep well because of the KATAHDIN SUMMIT TOMORROW!!

The climb up Katahdin ranged from easy to ARE YOU KIDDING ME?! The tablelands were a sweet reward for the walls we had to scale earlier. As we got closer to the summit they yelled for me to come into the group picture. I sat next to N Trovert and got tears in my eyes. Wow. My journey is over. It was pretty overwhelming. We did it!! OH MY GOSH! There were a lot of people up at the top, many of which I didn't know. I would say there were about 25-30 up there. I waited awhile to take my picture with the sign, and it was so cold and windy. I was eager to get somewhere not as cold. I took pictures with the Troverts and Beth and took pictures of the challenge coin

I brought up for Gypsy and a yellow golf tee that Andi VanHove had given me before I started. Wow.

I thought about those that are not with me that were unable to finish. Like Reaper and Alison and others that I know have gotten off the trail. Why was I special and was able to finish? It's just crazy.

We finished at 4:28pm. We took 4.5 hours up and 5 hours down. Nearly a 10 hour day. Crazy. It was an amazing day – hard, beautiful and emotional. Wow.

On the way home

Alison and I stayed in Maine until October 1st. Then we started the long drive home. It took us 2 weeks to get home. We visited family friends and trail friends in Nashua NH, Hanover PA, Harrisonburg VA, Greensboro NC, Clover, SC, Sylva NC, Knoxville TN, Corinth MS, Lewis KS, and a stop in Torrington WY for Alison to have a job interview. Then home to South Dakota. It was the most round about trip we could have taken, but it was so rewarding and great to decompress after the trail.

Hiking with my sister

There were many cool things about hiking with my sister. We don't have to pretend to get along and if we are mad we usually can tell. We already know just about everything about each other, so we don't have to make small talk. We were perfectly fine hiking in silence. But, the thing I liked the most is sometimes we would just sing. I love singing with my sister and being out in nature it is easy to burst into song to see how beautiful God's creation is.

I believe we got much closer while we hiked the trail. As I continued on by myself it was so nice to know that

Alison knew the people that I was talking about. It was amazing to have her meet me during her travel nursing assignment and that she could meet me in Baxter State Park the day before I summited. She cheered me on as I took off up the mountain and gave me a hug when I descended. It was so amazing that she was there, seeing the start, part of the middle, and then the ending. She knew what it was like to hike all day until her feet blistered. She knew what it was like to look at the trail and wonder "how in the world do I get up that?!" I loved sharing my AT hike with her and I'm so glad that she wasn't scared away when I sent her all those doomsday statistics. It was great to have a built-in hiking partner, someone I knew well. We have the ability to laugh at ourselves. We pushed each other. Once the trail gets ahold of you, it doesn't let you go. Even though Alison didn't complete the trail in 2012, she is planning on finishing the AT, just in sections. I hope to help her finish it and make more memories.

~ ~ ~

Renee currently lives in Alaska where she does as much adventuring as time allows. Life since the trail has led her to dabble in retail, house cleaning, working with young adults, and volunteering. She often visits Swan Lake Christian Camp. She raised money for the camp while walking the trail. www.myslcc.com

Renee also runs marathons, sings in the Anchorage Concert Chorus, and travels. Her most recent trip took her to New Zealand and Australia to do some hiking with Alison. Her experiences on the AT will forever be some of her fondest memories.

Alison lives in Colorado with her two cats and works as a labor and delivery nurse. Her recent adventure is studying to be a midwife. She aspires to finish the AT in sections.

Chapter 12 Living the Dream as Father and Son
Al and Shawn Olsavsky aka Big Dawg and Indiana Jones

When you hear about people hiking the Appalachian Trail (AT), most of the time they are either hiking it by themselves or hiking along with people that they met along the trail. However, there are some stories about groups of people planning and hiking the trail together, although it is much less common. The reason: each person has to commit 6 months of their time and dedicate it to the hike. In today's hectic world, it is hard to find the time! So, which groups of people have enough free time or open time to hike for 6 months? Two major groups are people who are retired from full-time work or people who are young and either finishing college or able to take time off from their education. Each member of our duo falls into these categories and we successfully hiked the Appalachian Trail - here is our story!

Al Olsavsky ("Big Dawg") here – a retired mechanical engineer and the main author of this chapter. My son Shawn is a full-time employee with undergraduate and graduate degrees from Purdue University. You might be wondering how we came up with the crazy idea of spending six months walking in the woods through 14 states. Growing up in the Washington, D.C. area, I spent a lot of time hiking and camping with the local Boy Scouts of America (BSA) troop. The AT was only a couple of hours away from our troop headquarters, so we enjoyed many hiking and camping trips to different sections of the trail. Although young at the time, my friends and I talked many times about someday hiking the whole thing.

Now fast forward about 30 years. When Shawn was 11 years old, we both joined Troop 103 in Indiana – he as a new Scout, I as an assistant Scoutmaster. Over the next 7 years, we went on many hiking, camping, and backpacking trips. We enjoyed the time spent in the outdoors and on many occasions I talked with him about the different places I hiked on the AT in my youth. Shawn became an Eagle Scout at a young age – an impressive achievement that we both share. His increasing age and maturity allowed us to go on the longer backpacking trips with the older scouts in the troop. One of our most impressive trips was a 65 mile hike in the Rockies of New Mexico at the BSA camp Philmont. I can point to this as another adventure that sparked our interest in hiking the AT. He remembered all these trips and my stories about the AT and lo and behold - the rest is history.

If we travel back to the year 2013, the situation was slightly different – I was still a retired engineer, but Shawn was working to finish up his undergraduate degree in aerospace engineering by December. He was planning on earning a master's degree at Purdue but his graduate studies did not begin until late August of 2014. On a side note, not that I am proud or anything, but he went all the way through college, including his masters, with a 4.0 grade average. During the summer of 2013, Shawn, after finalizing his schedule for the master's program, discovered that there was a large chunk of time that he now had free. Remembering all of our talk about hiking the Appalachian Trail, he realized that the time gap opened up the door of our amazing opportunity – hiking the entire AT together! He brought the idea up with me that summer and I realized that this may be the only opportunity that we would ever have to make the dream a reality. I knew that this trip would be a huge commitment both in terms of time and money. But for any parents reading this, how cool is it to be able to do something so incredible with a son or daughter? We finally agreed to go forward with the trip and the planning and research phase began!

Even after agreeing to the trip, I must say that I had several concerns and unanswered questions about the trail, ranging from personal safety to trip logistics and many other topics. Some of the concerns and questions were about things like the following:
1) Is the trail safe or do we need some kind of protection from people and animals (such as a gun, pepper spray, etc.)?
 2) If we get hurt in the middle of the woods, how will we get help?
 3) Since I am a bit older than my son, can I keep up with him? Can we actually hike 2185.3 miles?
4) Will we be able to hike about 15 miles per day in order to finish the trail before school starts?

We split up the different areas of planning the trip and performed research to make sure we were ready. We spent six months gathering data, answering questions, and creating a daily plan, including shelter stops, town stays, and food pickups. One of those steps included listening to a presentation given by a guy who hiked the AT in 2013. That guy was Jim Dashiell aka Funnybone, the driving force behind this book. We were only weeks away from beginning the hike when we met with Jim. He provided a lot of additional insight into preparing for the hike. By the end of the planning phase, we had acquired new equipment and dehydrated foods along with any other foods needed for the trip. (My wife packed boxes and sent them to planned stops).

We did practice hikes around southern Indiana that totaled up to about 200 miles before we left. I figured between the hiking and running five miles a day I at least had a chance to stay in the game. The practice hikes were useful in developing our stamina while also giving us an opportunity to test out new gear. One major change that we had to test out was using hammocks instead of tents. We made that change two weeks before we left. It was Indiana Jones's (Shawn) idea, and I admit I was skeptical about it at first. However, during our practice runs, the hammocks turned out to be a great idea! I would wake up in the morning and not have a sore back from lying on the ground – loving the hammocks!

Another problem to solve before we left was how to communicate with my wife and daughter. They wanted to be able to know where we are and that we were safe at the end of each day. Unfortunately, cell phone coverage in the middle of the woods is very situational and would not always be available to us. In searching for ideas online, I came across the solution to our problem – the SPOT. It is a one way GPS communications device that sends a location data point every 10 minutes. The data gets plotted on a custom webpage that is set up for each SPOT and any users with access to the site can see where we were and about seven days of where we had been. It uses the Google Maps interface, so you can zoom in to the area around our data point and see close up details. The SPOT also had three preset buttons on the device that you could customize with different notifications for the users back home. For example, we had a message that said, "Made it to the next stop ok. All is well." We sent that out every day when we were done hiking. The SPOT also has a "help" button that ties into the local emergency management system – you could still get help even while in the middle of the woods.

While were doing the planning, the word started to get out to family, friends, and neighbors. Most were saying how cool this was going to be with a father and son being able to share this opportunity together. 100% of our friends and family wanted to be kept informed of how our trip was going and where we were in the journey. Although the SPOT worked great, we were limited to sending emails or texts to 10 email addresses or phone numbers. With all of the friends and family that had heard about our trip, we immediately had well over 30 people who were tracking our journey and wanted to know how things were going. I told my hiker friend Todd about this and he said, "Why don't you configure Outlook to automatically forward the e-mail to everyone?" The tip worked like a charm. We were able to get the information to all of our friends and family!

Another thing we used was a hand held Garmin. It had a detailed map of the trail uploaded which showed shelters, parking lots, and details about the entire trail. It could also be used to find places to eat in towns, libraries, post offices, grocery stores, and everything else you can do, just like the one in your car. We used it mostly to determine how many miles we had left to hike or how far we had gone during the day. It was also useful for elevation info. We didn't really need it to follow the trail.

After all the preparation and planning, our start date, March 15, 2014, finally came. So we left Indiana and drove to Ellijay, Georgia to get ready for our adventure. To say the car was loaded would be an understatement. Besides the two of us going to hike, the car was filled with my wife, my daughter, one of her friends, our two cats, and lots of gear strapped to the roof. Seeing the gear on top of the car and all of us piling out of it when stopped was quite a sight to behold!

When we arrived in Georgia, it was a beautiful sunny day with temperatures around 70 degrees – very surprising since it was still winter! That weather did not last though – on our trail start date, we woke up to a cold, rainy day with temperatures in the low 40's. We picked our way carefully up a bumpy, muddy forest road to a parking lot with hiking access to the summit of Springer Mountain. We hiked to the summit of Springer Mountain in the rain and then returned to the car to get all of our gear. Now it was time to say goodbye to the family, which was quite difficult. Since it was raining, we said quick goodbyes so they could get back in the car to avoid getting soaked. It was a very emotional time for all parties involved. For my wife and daughter, we were leaving them and going on a trip with a lot of unknowns. It would be several months before we would see them again. My son and I were walking off into the woods in pouring rain -- not the type of day we wanted to start with. I remember thinking to myself as we left the parking lot, "What did I sign up for?"

My nightmare scenario for hiking and camping was being cold and wet. I hate it when the temp is in the 40's and raining because it is hard to keep the gear dry and to keep warm. This was our life for the first 4 days of the hike before we finally got a break in the weather -- no fun! At the end of the second or third day I was so cold and wet that I couldn't get warm. The hiker's paradox appears in this situation – even if you have good rain gear, you are still going to get wet with either rain water or sweat.

After starting on the trail, one of the first things we had to do was to establish our trail names. You can either get a trail name for yourself, have one given to you (which normally relates to something embarrassing someone saw you do), or keep your real name. Everyone on the trail has a trail name and real names are rarely used. I chose Big Dawg and Shawn chose Indiana Jones. Big Dawg is a name I got dubbed with in the past, but it also relates to me as the head of the household. Indiana Jones is appropriate too. If you remember the movies, the underlying theme is a father and son pair going together on an adventure to find some rare object. Although we are no Sean Connery (the elder) and Harrison Ford (the younger), we were about to embark on our own adventure into the unknown (and snakes – why did it have to be snakes?)

We realized early on that there were three unique experiences we would have on the trail. I will highlight them as they fit into my chapter:

1) Meeting many different people from all walks of life and with their own life stories.

2) People called trail angels who set up by the trail and provide food and drinks to the thru hikers.

3) Some great hostels and hostel owners.

On the first day we hiked about nine miles to the first shelter – a short day for the start of a long journey. At that shelter we met another thru hiker named Grey Ghost, a fantastic guy whom we liked immediately. At the time, we had no idea that we would follow each other way up the trail for many miles and become such good friends.

Grey Ghost lives in Roan Mountain, Tennessee. When we ran into Grey Ghost later on in Tennessee near Roan Mountain he asked us if we wanted to come and stay at his house since it was near to the trail. We thankfully accepted. His wife picked us up at the road crossing on Roan Mountain and drove us to the house, where we settled in and got cleaned up. The hospitality was so great -- a warm bed, a home cooked meal, and a dry place to stay. We even watched some television with them in the evening and just chatted. We still stay in touch with Grey Ghost today.

On the second day of hiking, we met a young couple from Michigan. They didn't have trail names yet, so we called them "Dallas" (the guy) and "Shorts" (the gal). When we met them on the trail, it was another one of those rainy and cold days. Somehow Shorts had gotten her hiking clothes wet and had to wear shorts to hike -- not exactly the best attire for a cold, wet day. Hence the trail name Shorts was born. Later that day, we went to Woody Gap (an unplanned stop) because I was wet and cold and needed to get warm and dry. We went to the Wolfpen Gap Country Store, a small hostel with bunks above a gas station. After we settled in, guess who showed up -- Dallas and Shorts! We talked for some time and they found out we were from Indiana. They immediately asked if we played Euchre -- everyone in the Midwest knows that game. So we spent the night playing Euchre, having a few beers and getting dry.

We also met a ridge runner named Razor. A ridge runner is a trail monitor who looks for maintenance needs on the trail and helps hikers as needed. Ridge runners maintain a certain stretch of the trail and walk it weekly to make sure everything is in good shape. We chatted briefly and went on our separate ways. A few days later we were in Hiawassee, Georgia at the gap looking for a ride into town. A lady in a large car pulled up and asked if we needed a ride. We declined politely as we had not yet figured out what we were going to do in town. A few minutes later, an older gentleman came out of the woods on the trail and went over to talk with the lady. But this was no stranger to us. It was Razor. He remembered us and asked if we needed a ride. With our plans firmed up and a familiar face, we gladly accepted. After they dropped us off in town, we said farewell and thanked them. We figured we would probably not see him again since his area of the trail was in Georgia. Boy were we wrong! Unbeknownst to us, we would meet up with him again hundreds of miles later. More on that in a later section. We offered to pay them for the ride. They said, "No way Jose!" This was one of our first experiences with trail magic – people being kind and generous, just wanting to help out expecting nothing in return. We would run into trail magic along the entire length of the AT.

It became very obvious from day one that people were intrigued by a father and son hiking together. On the trail and at home, we heard over and over how it was such a cool and unique opportunity. Indiana Jones and I had camped and hiked a lot before this in the BSA program and one thing we always enjoyed was a nice roaring campfire. It was a great way to unwind at the end of the day and a good place to swap stories. Starting day one, we made it a practice to build a nice fire every night, weather and location permitting. In the first couple of days, the only fires we saw were dinky little fires with hardly any flame. Granted, it was partially due to the wet weather, but it wasn't a true campfire! We quickly remedied that situation with great campfires wherever we stayed on the trail. I mention these two distinct, separate ideas together because as we moved up the trail, we became "trail famous" for these two reasons. Word on the trail has a way of spreading forward and backward. So when we met new people, they often had heard of us as either the father/son pair out doing the AT or those two guys who make great campfires! All of those interactions with hikers were great experiences and made the trail memorable. I was really proud of what we were doing and so was Indiana Jones.

We worked as a team in the early days and got very proficient at setting up and tearing down camp. Every day when we arrived at our destination, we had to put up the hammocks, cook dinner, clean up, get water refills, and gather wood for our excellent campfire. We knew what to do and if the work needed be done in a different order, we did so without missing a beat. Before we left Indiana, we had talked about how to plan the days so that we got to camp by five or six o'clock, which would allow for time to get set up at the shelters. We also wanted to leave time for a few hours of reflection on the events of the day, talk about the plan for next day, and visit with new friends. For me, that was the best time to really enjoy the experience and have some quality downtime with my son.

The first trail magic with food came at Unicoi Gap in Georgia. The trail goes right over a well-paved road and dirt parking lot on its way through the gap. As we were approaching the road, we could see that someone had set up a huge spread of things to eat and drink for the hikers. And everything there was free for thru hikers -- how cool is that?! After getting our fill of food and drinks, we headed up a steep trail to go over the next mountain – never around it, always directly over it. On the way out, I saw a hiker carrying a 12-pack of Bud Light! While nothing beats a good drink after a day of hiking, that was not very practical. We spent a lot of time trying to shed every ounce of weight to make our packs as light as possible. No doubt that hiker didn't keep that trend up for very long!

We met another hiker in the first two weeks of our adventure. A young lady named Peppa was hiking the trail by herself and she caught up with us just before we started climbing Tray Mountain. She was from New Hampshire and was familiar with hiking in the Whites, a series of steep mountains in New Hampshire that are part of the AT. After chatting with her for some time while resting, we took off from our stop point and immediately started up a steep hill. (Seems to be a reoccurring theme, doesn't it?) She and Indiana Jones were in power climbing mode. They blasted up the hill – Indiana Jones preferred not to stop while climbing up hills. I told them I would meet them at the top. We walked with her for a long way up the trail but since she hiked faster than us most days, we eventually lost her. We also met a young lady from Switzerland (Wildlife) who had also come over from Europe by herself but had teamed up with a young lady from North Carolina (Tatiana). Like Tatiana, we tried to adopt and befriend any Europeans when we met them. When you are in a different country, you don't know the ways, customs, and how to get things done in that place, especially if you have never been there before. Trying to learn a new place while trying to plan and hike the Appalachian Trail can be daunting. We met many Europeans from several different countries and every one of them was very friendly.

Over the next several days, we plowed through Georgia and entered North Carolina. The first climb of North Carolina said "Hello hikers – welcome to a lot of climbing!" Our welcome was a very steep, difficult rise up a mountain. Before we started that climb, we met a young guy who had come over from Germany by himself to hike the Appalachian Trail. We hit it off with him and we ended up following each other on and off all the way to Katahdin. After a few days, he ended up hiking with another German who we named Fritz. They hiked together until Fritz developed a stress fracture in his foot and had to get off the trail in Harpers Ferry.

On the way to the Smokies we had more trail magic, crossed the 100 mile mark of the trail at Mt. Albert, passed through Franklin, North Carolina, the Nantahala Outdoor Center, and Fontana Dam. As we made another huge climb out of Franklin, we got to about 5000 feet when the temperature dropped rapidly. Snow was falling and it was downright cold! Along the way we heard from other hikers that the temperature was supposed to drop down to single digits that night. Since our hammocks were not rated to temperatures much below 15 degrees, we got out the trail guide and searched for alternate solutions for the night. We decided to go to the hostel Aquone Hiker Lodge that was several miles off the trail. We called Wiggy and he came and picked us up at one of the road crossings. Wiggy was a great guy -- he was a part of the Special Forces in England and his wife was a personnel manager from a company in England. They came to the US 18 years earlier and built from scratch a lovely house and several cabins that serve as a hiker hostel in the winter and retreat for family vacations in the summer. By mutual agreement, Wiggy does all the outdoor stuff and Maggie handles the inside chores. Aquone Hiker Lodge turned out to be one of our favorite hostels on the trail. When you get there you have to take your boots off to go inside -- you heard me correctly. When you enter, you see why -- an absolutely spotless, beautiful place. They gave us clothes to wear while they washed our dirty hiking clothes and served excellent homemade meals that Maggie prepared. Wiggy was a great guy and an unusual character that intrigued us with his stories. We liked it so much that we stayed there for two nights!

Whenever we stayed in any of the towns or hostels I mention, we met up with other fellow hikers at night for dinner. The comradery and good times became even better as we began to learn about and bond with our fellow thru hikers. Our father/son trek was always a topic, but everyone had their own unique story to tell about why he or she was hiking the AT. Each introduction brought a new trail name into the myriad of names of people we met. I am glad that I had Indiana Jones with me – he has a pretty good memory. I usually forget names, so for a while I was always asking him, "Who is that person again?" I got better at remembering names the further down the trail we went.

We finally got to Fontana Dam, but not before we got soaked in a downpour! Luckily we were planning on stopping at the resort to pick up some food – we could dry out our gear there as well. The resort was only $59 for a room for two hikers and it was a very nice place. While it was not exactly a hostel, I think we would recommend it to anyone since it was one of the best deals on the trail. The resort had food supplies at a store, a dining room for meals, a laundry area for the hikers, and a post office. It had everything that we needed! We left the hotel the next day to cross Fontana Dam (the trail goes right over it) and enter the Smokies. This was a very emotional moment for us. The Smokies are one of the must-see areas of the AT. Being able to hike it with my son was great and it was an experience we will not forget! The storm that chased us into the resort was rain at the lower elevation but turned into snow in the Smokies. I knew early on that getting through the Smokies would be tricky. The snow definitely added to the degree of difficulty. The first night we finally got to one of the shelters after a tough 4,000 foot climb. When we got there, we looked inside the shelter and there must have been 25 or more people standing inside – definitely no room for us! In the Smokies, you aren't allowed to camp outside the shelter. So we went to the trail guide and looked for other options. Indiana Jones was always on the ball with finding ways to overcome issues on the trail. It turned out there was another shelter about 3 miles away. I was a bit skeptical about whether there would be space in that shelter, but we pushed on and were pleasantly surprised when we arrived. There were only a couple of people there and plenty of space for us. Problem solved -- good call Indiana Jones!

In the Smokies, we met our newest set of hiker friends, a young guy and his girlfriend who came all the way from Switzerland! I have to say they were some of the nicest folks we met on the trail. We spent many nights in camp with them all the way up the trail, even reaching the Katahdin summit on the same day. They didn't have trail names when we met them, but that was soon remedied with Chocolate and Cheese – since they are both Swiss. We stay in touch with them to this day.

Some of the best aspects of experiencing the trail were the views and scenery along the way. We saw mountain views, unique shelters, river crossings, bridges, rugged scenery, and many things that you can't see from the inside of a car. These places can only be seen by walking to them, and they are very rewarding when seen after hiking many miles. We often talked about how we were probably seeing things that very few people see, especially the further away from town we got. It was definitely one of the most amazing parts of the hike, and I am glad that I got to share it with my son.

The Smokies have lots of scenic places, such as Rocky Top (made famous in the song "Rocky Top"), Clingmans Dome (just 50 feet short of the highest point in the eastern US), Charlies Bunion (a cool rock formation), and the mountain town of Gatlinburg. The trail up to Clingmans Dome is extremely difficult -- you are at 6,000 feet elevation and it is very steep. I could only do a couple hundred feet at a time, especially with the melting snow making the trail slippery. I think Indiana Jones was getting used to me having to slow down on the steep stuff. We finally made the top and were treated to a majestic 360 degree view on a crystal clear day. I had been here before, but it was really neat sharing that spot with my son.

After soaking in the views and resting, we went on down the trail where we met up with friends Todd and Kelly who drove all the way down to Newfound Gap from Indiana to pick us up and spend some time with us in Gatlinburg. How cool is that?! This was the first place we actually took a zero (rest) day after 17 days straight of over 12 miles per day. We periodically took rest days throughout the trip to give our bodies opportunities to rest without hiking every day.

After catching up with our friends and enjoying the rest, we returned to the gap and kept going north. The next 40 miles of the Smokies was my favorite part of the trail. You walk along a 5,500 foot high ridge for over 20 miles with steep drop-offs on both sides. It was an amazing sight. On the way out of the Smokies, we crossed a point on the trail that we had hiked before. Years earlier, Indiana Jones and I went on a summer adventure with the Boy Scout troop and we stayed at Cosby campground and hiked the access trail to the AT. When we got to that point on our AT journey, it was pretty cool – it was something Indiana Jones and I had shared in the past and were now revisiting it on another adventure!

After the Smokies, we crossed through a big field called Max Patch – a grassy hill that is also known as a bald. Max Patch is a popular spot where hikers and locals congregate to visit and relax, especially during weekends. When we got there, it was a cool day with gusty winds and luckily some sunshine. Even with the cold weather, word on the trail was that there was trail magic at the top of Max Patch. With extra motivation, we quickly made it to the top of the bald.

It is worth mentioning at this point what Indiana Jones had been doing every night on the trail since we left. At the end of each day, he would write down the day's events on his Kindle. When we got into the next town, he uploaded all the entries to a website and added more as we moved on. We would also send out an update email each week to let people know how the journey was going. These activities normally only occurred in towns because that is the only place we had Internet access. I don't know where he got the energy to do the log after hiking each day.

Let's return to Max Patch. As we were walking along Max Patch, I heard a female voice calling my name. I was absolutely shocked and confused -- how would anyone around here know my real name? I thought it was a mistake, but then I heard it again. I finally noticed someone coming towards us and waving us down. But since it was cold everyone was bundled up and I still couldn't tell who it was. Finally, I figured out it was Jan from our home town of Cicero, Indiana. But I was baffled as to how she could have possibly known where we would be at this exact moment. Well, Indiana Jones finally caved in and explained that he had been emailing back and forth with her about our progress. Jan and her husband Larry had been driving around the area on their vacation and stopped by to meet up with us. They brought us some goodies and we visited for a couple hours before heading up to the next shelter. What a great and unexpected experience!

From Max Patch, we went on to pass through Hot Springs, North Carolina and Erwin, Tennessee, both really cool trail towns. In Erwin we stayed at Uncle Johnnies Hostel on the Nolichucky River, another unique trail hostel. We had planned a short hike from the previous night's shelter so we could spend some time exploring Erwin. Shortly after leaving Erwin we were off to hike up Roan Mountain. I mentioned earlier that we had met up with Grey Ghost several times on the trail. He invited us to visit his house and stay overnight. We thankfully accepted and had a wonderful stay there with home cooked meals, nice beds, and good company. Thanks Ghost, we really enjoyed it!

Little did we know that the nice time from the previous evening would quickly turn into a very difficult situation. The next morning, Becky dropped us off at the highway crossing to continue on our AT journey. The skies were dark and looked like rain but we were not too concerned – we had seen dark skies before. What we didn't consider was the upcoming terrain. The next ten miles took us over a series of balds, one after the other after the other at elevations around 5,500 feet. The balds are exposed, open grasslands with few, if any, trees. As we left the road crossing it started to rain. No big deal – we just put on our rain gear and just kept walking. Unfortunately, the weather was not going to cooperate with us. By the time we reached a barn shelter to stop and eat lunch, the weather was turning worse. It had transformed from cool temperatures and rain to freezing temperatures, high winds, and sleet.

Indiana Jones and I looked at each other and paused a moment – trying to figure out if we should keep going forward or return to the road and call Becky. The temperature was dropping rapidly and the rain had become sleet, but we thought that as long as it stays sleet or snow, we would be able to make it to the next hostel. Rain and cold temperatures would have been a no-go situation. Doing the manly thing, we decided to go forward and pressed on into the weather. An intense winter storm was about to unleash itself on us with winds up to around 40 miles per hour. Snow and sleet pounded us. I have been in many cold situations before, but my fingers were numb all the way down to the knuckles. I don't think that has ever happened to me before or since that day. With no trees, the blowing wind made going over the balds tough. The wind was blowing so hard that it kept blowing me off the trail. Finally, Indiana Jones saved the day and walked in front of me to provide a wind break. We knew if we could get through the balds, we would drop down into trees and continue descending into the next town. By the time we crossed the 10 mile stretch of balds, we both had ice coating all the way up and down the left side. That was our worst weather day on the trail.

After making it out alive, we stayed at the Mountain Harbour Hostel. If you hike the Appalachian Trail, don't miss this place! You can get really nice rooms inside or spend less to sleep out in the barn. The breakfast was amazing. There were two great big tables full of every kind of breakfast dish you could imagine, and all for $10. We met and chatted with the owners. The people who owned the place were great people and gracious hosts. It was definitely a good place to get warm and dry.

Soon we were in Virginia, the state with the longest section of the AT -- over 500 miles. A few days later, we were in Damascus -- a trail town for sure, since the trail goes right through the heart of the town. There were a couple of hostels and some good places to eat. Earlier, we had met a young German couple on the trail, Robert and Svenja. When we stopped to talk to them, Robert offered us one of the beers that he was carrying. I guess that figures as most Germans love their beer. I drank one, but the real story was he had a few in the pack and wanted to lighten the load! It was still very nice of him to offer to share one with strangers. After running in to them again in Damascus, we went to the grocery store and bought a 12-pack and put it into a bag along with a dozen of those

red plastic cups you use on a picnic. Seems to be a strange
purchase – why do we need those red Solo cups? Well,
Damascus is a strange town that has a law stating you
cannot be in public with an open beer container.
However, the workaround to the rule is that if you pour it
into an unmarked plastic cup, it is acceptable. So we took
the bag, the beer, and the cups and went to find Robert
and Svenja to invite them to drink beer with us in the
park. We had a great time and really enjoyed getting to
know these folks.

That story is an example of how the trail works and why
you meet so many great folks on the trail. People come
from all walks of life and are brought together with their
fellow hikers for one purpose -- to hike the AT. Indiana
Jones and I talked about this a lot and how the great
people we met really made the experience more
enjoyable. Back in North Carolina, we had one night
around the campfire that showed some of the diversity of
the trail. Around the fire, we had Indiana Jones and
myself from Indiana, a guy named Shroomer from
California, Wildlife from Switzerland and Tatiana from
North Carolina, Fritz and Pyro from Germany, two young
friends hiking the trail together (Rain Pants and Mio), and
several others. We were all there with something in
common and getting along with each other like we'd been
friends all our lives.

North of Damascus, we hiked through Grayson Highlands
and came across a herd of wild ponies. They would come
right up to you and try to eat the salt that accumulated on
you from sweating. That was a truly unique place and
experience that I was glad to share with Indiana Jones.
We went through many small towns in southern Virginia.
The weather finally started turning fairly nice. As we
moved north we stayed a night at a place called Dismal
Falls (nothing dismal about it – the falls were great.) On
the way to the campsite, we saw a man and woman

walking toward us. As they got closer, I thought that they looked familiar. Remember Razor, the ridge runner from Georgia? The couple turned out to be him and his wife! They were walking the trail back toward Damascus for the big trail festival. That was hundreds of miles from where we last saw them! It is amazing how the trail continually brings people together, especially people we had not seen in several weeks.

We had the campsite to ourselves and camped along the stream and falls. This was one of several nights when it was just Indiana Jones and me; no one else around. It gave us a chance to talk one on one. It is hard to make time for father/son talks like this in your busy everyday life. The next day we went to Woods Hole Hostel and it is well known as a great place to stay. It is run by Neville and her husband Michael. They provided home cooked breakfast and dinner, unbelievable! You could stay in nice rooms or less decorated ones in the barn – we decided to stay inside this time. Great people here and we really enjoyed the stay. The next stop was Pearisburg. Indiana Jones knew a young lady named Amber who was attending Virginia Tech, which is just down the road in Blacksburg. She gave us a tour around the campus. We took her to lunch and she dropped us off at the hostel in Pearisburg. Thanks Amber!

We stayed the next night in the yard of a private individual's home called the Captain's. He let hikers use his yard to camp and had a fully stocked refrigerator on his back porch with all kinds of drinks for the guests. The only interesting thing was that in order to get there, you had to sort of use a "manual zip line" to cross a fairly wide creek on a metal cable and swing set seat. Indiana Jones thought it was great, I didn't share his enthusiasm. But we made it and had a good night's rest.

Traveling on, we arrived in the Catawba/Daleville area of Virginia. The first stop was the Four Pines Hostel. While this was not the nicest place we stayed on the trail, it was arguably one of the most fun. Our stay there came at the end of a tough day, where we slayed the dragon, known on the trail as Dragon's Tooth. It was a hot day with a tough climb up and an impossibly steep descent on the other side of the Dragon's Tooth. However, we were rewarded with amazing views from the top of the Tooth and a great opportunity to chill out at the hostel. The Tooth is a place I will never forget, especially sharing the views with Indiana Jones. At the Four Pines Hostel, we met up with a young English couple named Harry and Fran who had come over from the UK to hike the trail together. We got along great with them and they were very friendly and funny. The owner of the hostel, Joe, enjoyed having fun as much as we did. About 5 o'clock, he told everyone to get into the truck. We drove into town, picked up a pizza and some beer, and went back to his place. He had set up a place to play the popular tailgating game called corn hole (or bean bags, bag toss, etc). He had chairs all along his garage, so we ate pizza, drank beer, and watched and played corn hole all night. The best part was watching Harry and Fran play the game -- there is no such thing as corn hole In the UK.

The next day we were going up to McAfee Knob. Unfortunately, when we got up there, Indiana Jones was very sick. There are many things to watch out for on the trail, including diseases from rodent droppings and other stuff. We weren't real sure what he had, but he was very sick. We decided to see if he could do any hiking at all. It was only a few miles to the highway access point to McAfee. We figured if he couldn't do it, we had a convenient pickup point. Once we got to the road, it was clear that whatever was bothering Indiana Jones, he had not improved. We got a ride to Daleville, stayed at a Howard Johnson's overnight and got a ride back the next day. While Indiana Jones was resting and recovering, I met up with Chocolate and Cheese and ate dinner with them. The next day, Indiana Jones was doing much better, but since we had a 20 mile day planned, we decided to slack pack just in case something came up. Slack packing means you leave the big pack in town, bring only what you need for the day, then walk back to where you left your pack. That turned out to be a great solution. McAfee Knob is one of the most photographed spots on the trail. Of course, Indiana Jones had to go sit out on the edge of it and get his picture taken.

As we continued hiking, we had planned a special reunion with the family. My wife and daughter drove from Indiana to Buena Vista, Virginia to join us right after we crossed the longest bridge on the trail, which goes over the James River. It had been two months since we left them in Georgia, so you can imagine how good it was to be reunited. Of course, they adopted trail names too, so "Hello Kitty" and "Trail Momma" did their first time on the AT. The next day we planned a hike with Hello Kitty. Trail Momma set up a trail magic station at the road crossing where we would stop that day. We decided to do a slack pack day so we could enjoy the time with Hello Kitty. On the way to the trail, we drove by Pyro and Fritz. They hopped in the car with us and we all hiked together. In the meantime, Trail Momma provided trail magic at the other end to other hikers following the Appalachian Trail. The slack packing was great and the trail magic rewarding after a long day!

Hello Kitty is Indiana Jones's twin sister. She loves the outdoors and could hardly wait to do some hiking with us. In fact, between this visit and some others visits she did later in the trail, she walked about 80 miles of the AT. I am sure she would have done the whole thing with us, but the window of opportunity was not there like it was for Indiana Jones. In fact, she was right in the middle of her second year at Purdue's Veterinary College. Purdue is known for its tough and demanding curriculum. So now, not that I am proud, of course, she has earned the title of Doctor of Veterinary Medicine, and is now working on her internship and residency.

It wasn't too long after their visit that we entered the Shenandoah National Park. At Shenandoah NP, we immediately ran into three Marines doing the Shenandoah section of the AT with the cutest beagle we had ever seen named Beans. They were great guys and we hit it off with them right away. We hiked the entire 105 miles with them. We also saw our first bear of the AT -- not just one but six, all within about 24 hours. It was pretty intimidating the first time we came across the black bears in the wild near the trail. We did not bother them and they did not bother us. All of the wild animals in the Shenandoah NP were unafraid of people. They would come very close to you without being scared.

While most of the hike was great for us, there were some times during the trip when things were not so rosy. These small things became more of an issue as we reached the 1000 mile mark at Harpers Ferry. It would usually come at the end of a long day or sometimes earlier when we were really putting out effort. I would get tired and worn out and on particularly tough days, Indiana Jones would be tired too. After the 15th or 18th mile of the day, all I wanted to do was get to the next shelter. When you are really fatigued, you can get kind of rough around the edges and your normally pleasant demeanor does a 180 degree switch. That happened to us sometimes, especially when I was really worn out.

As we made our way to Harpers Ferry, we had to go through a section of trail known as the Roller Coaster. It is 14 miles long and is a series of up and down elevation changes of about 600 feet. Initially, I thought that wouldn't be a problem since it seemed to be peanuts compared to what we had done up to this point. Well, we discovered some things we didn't anticipate. The trail was total rocks and boulders. The daytime temperature kicked in and it was hot. And on top of all of that, the trail was taking its toll on me. I was finding it harder and harder to rejuvenate each day. I was discovering that I'd had a major misconception about the trail experience. I believed that the training before the trail and from hiking on the trail every day would ultimately make me better conditioned to handle the strenuous daily activity. That did not end up being the case. The strenuous days accumulated and wore on me quicker than my body could adjust to it.

The struggle was particularly difficult on the day we were trying to get off the Roller Coaster. We found a way to get through it and kept moving on. When we finally got to Harpers Ferry the next day, I was really tired. Luckily that was a zero day for us so we could regain some energy. My cousin Dale, who I hadn't seen in a while, came up from Washington, D.C. and spent the day with us. We enjoyed some of Harpers Ferry cuisine and took in the historic sites. Harpers Ferry is like walking into a history book -- so much took place there during the Civil War. We checked into the Appalachian Trail Conservancy headquarters and got our names registered. We were thru hikers number 269 and 270 for 2014. That seemed really cool to me as Indiana Jones and I were now halfway and that record will be there for everyone to see. It was a great place to rest, and we appreciated the visit Dale!

Although we had reached the mental halfway point in Harpers Ferry, we still had about 80 miles to go until we reached to the true halfway point of the Appalachian Trail. Since the trail changes in length slightly from year to year based on reroutes and trail damage, the actual halfway marker is temporary since that can easily be moved to the new location. I wonder who goes out and actually measures the new halfway point each year. Do they have to walk the entire thing to do so?

A little further on, we came to a place where I spent many summers when I was a Boy Scout in Maryland. We would normally have our week-long summer camps at Pine Grove Furnace. It is a nice state park in Pennsylvania with a lake. I remember going to the little store for ice cream once that week. Well, wouldn't you know, the store is still there! It is the custom of thru hikers to celebrate making it to the halfway point of the AT by eating a quart of ice cream. Now, this is not just any ice cream, but Hershey's ice cream. We excitedly took on the Half Gallon Challenge. Next door to the store is the Ironmasters Mansion Hostel. After eating our ice cream we didn't wanting to leave the camp. I know the building was there when I was a Boy Scout, but I am not sure it was a hostel. I had a good time telling Indiana Jones about my memories of this Pine Grove Furnace. He could relate as he and I have been to many summer camps in Indiana. I also found the beach on the lake where I learned to swim and where I did the mile swim test the second year I was there.

There were a lot of things I told Indiana Jones about Pennsylvania. Hershey Park is a place I went to as a child. The trail comes within a few miles of it but not close enough to easily visit. I had a great aunt and uncle that lived in Pine Grove, maybe three miles off the trail. My grandparents lived in Schuylkill Haven, which was not far from the trail. They had a summer home that was only a few miles off the trail. I was born in Pottsville, also not far off the trail. We had a zero day in Hamburg, Pennsylvania, also a place I had visited many times before. I remember a place called Roadside America that I thought was close by. It is a model train display that is at least 200 feet long and 100 feet wide and is part of a museum. We found out it was just a couple of miles away, so we went over there for the afternoon. It is exactly as it was when I was a kid and the train table depicts the surrounding area, so it is a great conversation piece. The world's largest Cabela is in Hamburg, which we had to check out, of course.

As we made our way through Pennsylvania, the trail was easier to traverse except in the areas where the trail is covered in the infamous rocks of "Rocksylvania". Those areas are very difficult and tedious to hike, but we took our time and navigated the maze of rocks. There are several sections of the trail that are purely rock – you can't see the dirt trail beneath all of the rocks! But it was not the whole state. Pennsylvania also has many stretches of open field and farmland, allowing us to hike 25.6 miles in one day, our longest mileage day together on the AT. I remember a song called "Nothing's Gonna Stop Us Now" playing on the iPod that Indiana Jones had attached to his pack for the times we needed music. I had the greatest feeling hearing that song because we were having a great day and I had recovered from some of the lack of energy I had earlier. We were on our way to the next state and we really felt like we were all set to finish the whole thing.

We pushed on into New Jersey. The terrain was less strenuous than most other places but still very challenging. We saw the only other bear of the trip here -- it was a big boy! New Jersey is said to have the largest bear population on the trail. So it was surprising that we only saw one during our time in the state. In Vernon, New Jersey we stayed in the basement of a church hostel. Earlier that day, we met another father and son team, Kendan and Texas Slick. They were about the same age as us and we hit it off with them at the hostel. When I asked them where they were from, I couldn't believe the answer. They told us they were from Carmel, Indiana -- that is only 15 minutes from where we live! What are the chances of that happening? At dinner that night Indiana Jones and I discovered something else about their hike. As we listened to them talk, some of the struggles they had were exactly the same as ours as it relates to the father/son relationship. Indiana Jones and I talked about it the next day and laughed. At least our struggles at times were not way out in right field -- it was the same kind of stuff.

It wasn't long before we reached New York. Indiana Jones and I both needed some down time and did some slack packing around Greenwood Lake. Then we pressed on toward the Highland Falls area where we climbed Bear Mountain, a very popular area for the locals. We ate lunch at the top of the mountain. Then, we descended thousands of steps down to the lowest point on the AT. The lowest point of the AT just happens to be in a zoo. That's right – a zoo. It was just a small zoo, but it was very busy that day and it was a unique thing to see on the AT. At the exit of the zoo, we crossed over the Hudson River. We met up with Trail Momma and Hello Kitty again and had a joyous reunion. It was great to see them again and to spend some time together hiking and resting.

As we drove to the hotel that night we noticed a sign for "West Point". I thought it referred to some feature on the river. It turned out it was the real academy! We toured it the next day and it was really awesome. Since we had a few days together as a family, the plan was for Hello Kitty to hike with us the first day while Trail Momma set up another trail magic station. Then the next day they would trade places and do the opposite. The first day was long, another 20 miles, and we ended up camping at Fahnestock State Park. It went well but it was long and tiring and the campground was only so-so.

The next day, it was Trail Momma's turn to hike with us. We planned it that way since the second day would be a little easier, but still fairly long. Unfortunately, the easy day did not turn out to be so easy. There really is no "easy" trail on the AT, just less hard, hard, and harder. The trail was very rough, very rocky, and we had to climb around rocks and boulders in some spots. We had not walked very far on the trail when Trail Momma slipped on a rock and went to the ground. A few cuts and scratches later we were on our way again. Later in the day, we were walking on some very difficult rocks. Once again she slipped and fell, causing a number of cuts, bruises, and scratches to her face and legs. She had both hips replaced just a year or so before our trip and was not back to 100%. After the second fall we decided we better get her off of the trail. Indiana Jones took the GPS and looked for the nearest road. He got the coordinates for that point and then called Hello Kitty to give her the coordinates. About 30 minutes later we arrived at the road crossing and she picked up Trail Momma. They went to the next shelter and we met them there for the night. We did a couple more days with Hello Kitty hiking and Trail Momma doing trail magic.

We soon strolled on into Connecticut, ready to tackle another state. When we met with Funnybone before the hike, he told us to make sure to stay at the hostel run by Maria McCabe in Salisbury, Connecticut. So we made reservations and started toward the town. When we got to Maria's place, it was reasonably early and we were greeted by two other hikers, Don's Brother and Purple Pants. Don's Brother hiked the trail in 2013 in memory of his brother. He also wrote a book about the trail called *Don's Brother*. Purple Pants had caught Lyme disease and was headed home. This was our first look at the telltale red circular marks on the skin from the tick bite.

Maria is in her 80's and has been running the hostel for 15 years. She was a very interesting person, a gracious host, loved to tell and listen to trail stories, and she provided us with a very nice room for the night. After we met her, Don's Brother took us into town and we ate dinner and got food provisions and some beer. We returned to spend the evening with everyone and enjoyed having a few beers, telling stories, and had a few laughs. Funnybone was right; this was a good place to stay the night!

I should mention here that the rest of our story has some very emotional moments for both of us. I think you will get the picture as I describe the events as they unfolded.

It only takes a few days to cruise through Massachusetts. One day we stayed at the Berkshire Lakeside Lodge which is just outside Lee, Massachusetts. That area was really cool and the town of Lee was neat. Indiana Jones and I went to a nice place in town to eat. Then we got the food supplies we needed and went back to the hotel. The following day was one of the few hot days we actually had on the trail, and the next day was going to be in the 90's. We were facing a steep climb and it would prove to be very tough. As we started up the hill around 8:00 a.m. it was already around 85 degrees. I was really struggling to get up this mountain. Then the heat really kicked in and made it even more difficult to climb. We always kept some sort of chocolate candy in our lap belts in order to provide extra energy while hiking. I started slamming those down to try to get a burst of energy. We finally got up the hill and were able to take some rest. While I was eating more candy, I noticed Indiana Jones wasn't saying much, so I asked him what he was thinking. He answered, "I am trying to figure out what we can do to help you keep going on the trail." I wasn't too surprised as I was having a tough time. That was only the start of what would be a very tough day.

Late in the afternoon we were almost to our stop for the night in Dalton, Massachusetts. About 2 miles out from town, we stopped at a shelter to adjust our gear and to take a bathroom break. I was feeling pretty good because the day went much smoother after such a rough start. At the shelter, I was urinating and got a real shock – my urine was coming out blood red! I immediately thought of worst-case scenarios – cancer, kidney problems, etc. Whatever it was, it could be a serious problem. I said to Indiana Jones, "I have a problem and we need to get to a hospital." I explained what had happened and I remember the look on his face. We got into Dalton and there was not much help there -- no hospital or emergency care. We called my wife's brother, who is a doctor, and he thought it was probably a kidney stone that may have passed or could still be in there. We were going to be near a hospital in Williamstown, Massachusetts in two days. We decided I needed to be checked out as soon as possible. We were very near Vermont and about to enter some very rugged remote areas. I did not want to be in the middle of the woods and have a stone to pass.

A few days before that, the right side of my face started not responding to my facial muscles. My eye sagged, my mouth sagged, and I had no control on the right side of my face – it was basically nonfunctional. I wasn't too worried about that as some minor things like that had happened before. But the hikers were looking at me like I was weird because my speech was impaired.

We finally got to Williamston and tried to find a doctor, but no luck there. We arrived in town on Saturday, July 4th. After getting settled in town, we took a cab to North Adams emergency room. The bleeding while urinating had stopped but my face was still not functioning properly. This was a big moment for both of us – we needed to know what was going on with me. After 1,600 miles with no issues, would I have to get off the trail and leave Indiana Jones to go on his own? Did I have something serious?

We didn't have to wait too long. When the nurse at the hospital saw me come in, she called the doctor before I got to her desk. She interpreted the face sagging as a stroke. We told the doctor all of the issues. The doctor did a CAT scan to verify if there was a kidney stone and blood tests to diagnose the facial problem. After quite some time, the doc came in and said that the CAT scan did not show a lodged stone and that it most likely passed. The doctor thought the face sagging was a condition called Bell's Palsy. I had never heard of that! The interesting part is that it is usually caused by Lyme disease, which results from a tick bite. The blood test confirmed I had Lyme disease, although another blood test needed to be performed to confirm the presence of Lyme disease. The treatment was medication to keep the eye moist and it should clear up in a week or so. All of that came as a big relief to both of us because it wasn't going to prevent me from continuing the hike. Back to the hotel and a couple of celebratory beers and we called it a day. I could only imagine what was going through my son's mind as all of this was playing out.

The good news only lasted a short week. We were now in Vermont and Vermont is known for two things – roots and mud. The roots all over the trail look like the mangrove roots in Florida and are elevated about six inches above the trail. Vermont also has mud bogs, muddy trails, rocks, and of course, steep climbs. We were pushing towards Manchester Center. The trail seemed so crazy there and I was having a day where I was about fed up with all the rocks and roots! Indiana Jones and I were having a little disagreement over the topic. We were standing still when my foot slipped off of one of those roots and rolled sideways. I knew immediately I had a big problem. I felt a painful tear in my right foot. We decided to get to the next road, which fortunately was only four miles away, and called for a ride to Manchester Center,

Vermont. I had a feeling this was not going to be something that would heal quickly.

We went to the hospital in Bennington the next day. The doctor X-rayed and analyzed the foot. There is a tendon attached from under the foot to the side and he said he could feel it was torn. Then the doctor said something that I dreaded hearing, "You need to get off the trail and let it heal." I broke the news to Indiana Jones and we went back to Manchester Center. Up until the past week we had hiked 1,600 miles with no injuries or sickness, except for the one day with Indiana Jones, and then all this stuff happened in just a few days.

We were staying at the Green Mountain House Hostel in Manchester Center, Vermont. This was another great place to stay -- Jeff the owner is great and we really connected with him. During our down time, we had to plan our next steps for the trip. We were about ten days from the White Mountains. I had looked forward to seeing the Whites the whole trail and I did not want to miss it. So we planned to meet back up in Glencliff, New Hampshire ten days later and pick up there. I had to hope that ten days was long enough to heal my foot, but I knew that was pushing it. The next day, Indiana Jones caught his ride to the trail and Jeff took me to the bus station. I went to Rutland and took the train home. Of course, saying goodbye was tough, but we did the man thing and kept it light even though it was very emotional.

I will tell you as a father I was concerned for Indiana Jones and thought about all the stuff that could go wrong. If this had been the beginning of our adventure I would have been terrified to leave him. However, these four things made it better: 1) we had hiked for over 1,600 miles, 2) we knew how the trail worked, 3) there would be a group of 25 or so hikers around or near him that we already knew, and 4) he would have the SPOT for a real emergency. I went home but I stayed close. I tracked his progress every day, hour by hour using the SPOT and cell phone. I also took care of arrangements for his stays at places along the trail until I could get back.

While at home, I also plotted out a plan to do as much slack packing as possible in the Whites to make sure the foot didn't get torn up again. Finally the day came to go back - I was so excited. I took the train to Springfield and stayed the night with some relatives on my wife's side. I had not met them before, but they were so nice to help me out. From there, I caught a bus to Hanover and then I called a shuttle person to get me to Glencliff. When I finally got there late in the day, Indiana Jones was waiting with a smile on his face!

We did some final planning and off we went the next day. We got a ride to the first point and slacked back to the hostel. The sign at the trail head read, "This trail is extremely tough, if you lack experience don't do it." I thought, well I guess we will find out about the foot real quick. I don't know how to describe the trail in the Whites and on into Maine. We used both hands and feet to climb the trail and descend after the peaks. This trail was crazy difficult.

So we moved our way through the Whites. If you haven't been there before, you should plan to go. That was one of the coolest places on the trail. The notable things we saw were Franconia Ridge, Mount Washington, and the AMC Huts that are available along the trail. Franconia Ridge is way above the tree line and is a ridge that is very narrow with steep long drops on either side. These all provided awesome views of the valleys and lakes below. Mount Washington is like no other place on earth. It holds the record for the highest ever wind speed at 241 miles per hour! It is 2,000 feet above the tree line and the views are fantastic. To get up to the summit, we could either hike or take the Cog Train. We took the train twice and slack packed the AT in both the northern and southern directions. For me, this area was one of the pinnacles of the trip. Being able to see it with Indiana Jones was priceless. I have a picture of us on my phone standing at the top on July 26th of 2014. It was only about 40 degrees and the wind was howling as usual, but awesome!

We pushed on through the Whites, some days hiking with the pack and some days slack packing. Toward the end of the Whites, we stayed at the White Mountains Hiker Hostel. The owner, Marni, does a fabulous job of giving the hikers what they need and maintains a very neat and clean place. She made a breakfast to die for every day. She provided us with transportation on the slack pack days. Don't miss this place if you hike the trail in the Gorham, New Hampshire area.

It was finally time to leave and finish the hike, but we had the task of getting through Maine first. Gorham is right on the border, so we got into Maine quickly. We went back to carrying full packs and the first day hiking was tough. Steep trail, big climb, really taking its toll on me, but the foot was hanging in there. We continued a few days after that with few problems. Then came the day to go through the famed Mahoosuc Notch -- I knew this would be tough. Funnybone told us all about it before the trip. It carries the reputation as "the toughest one mile on the trail." Considering all the tough terrain that we had tackled before the notch, how hard could it be? Well, it definitely lived up to its name. It was a 2500 foot descent via steep rocky trails just to get to the notch, and then one mile of climbing over and under and around and any way we could to get through the notch. It took us about 2.5 hours just to go one mile. If you don't believe me, just search on Youtube and watch the video. The rest of the trip that day was hard as well. We hiked 2,500 feet back up steep rocky trails and then 3,000 feet down to Grafton Notch and the highway on steep rocky trails.

Well, as (bad) luck would have it, I re-injured the right foot somewhere in the notch. I really don't know where. I told Indiana Jones about it after we exited the notch. The problem was we still had the steep climb and descent to get out of the woods, a mere 7 miles! So on a pretty sore foot, we climbed up and then went down. By the time I got to the pickup area, I could hardly walk. I didn't want to admit it but I knew that was the end for me.

We stayed at the Pine Ellis Hostel in Andover, New Hampshire. It was a nice, friendly place with good accommodations. While I waited to see if my foot had a chance of recovery, Indiana Jones slack packed a few days. Unfortunately, it was not meant to be. I could hardly walk on it and it was time to decide where we go next. We decided that I would go home and he would go on – there was no other choice after coming so far. So once again we did the man thing and kept it light on departure day even though it was pretty emotional.

I was no longer worried about him on the trail. We knew how it works; he would be around others that he had been with for 4.5 months. But the one thing I did worry about was stream and river crossings. In Maine there are about 12 of them. The difference is there is no bridge -- you have to walk through the water. Most of the time it is ok, but the weather is quite volatile and could change a simple task into a very dangerous one. He would also have to go through the 100 Mile Wilderness. No services, only two dirt roads, and nothing else that looks like civilization – pure wilderness. He would be out of cell phone coverage while in the wilderness.

I made it home easily and went back to watching his progress every day on the SPOT website and setting up reservations at the various places he would stay. We were able to talk on the phone and take care of business up until Monson. That is the start of the 100 Mile Wilderness. Our plan had been to stay at the 100 Mile Wilderness hostel in Monson, Maine. I contacted the owner Phil and I explained the situation and asked him to watch out for my son and make sure everything was cool. We also had Phil do a food drop halfway through the wilderness. Phil came through and did everything I asked. I talked to him on the phone a couple of times and felt like I'd known him forever even though we had never met. Thanks Phil, I really appreciated you help!

Just as Indiana Jones was leaving Monson there was a huge storm coming up the coast. It dumped 12 inches of rain on New York and ultimately several inches of rain in Maine. This was my worst nightmare as Indiana Jones still had many river and stream crossings. I hoped he got my texts and messages before he left. Cell phone and internet were not an option once he was in the woods. The storm hit Maine a few days after he started the 100 Mile Wilderness, which normally takes about five or six days to hike through. I followed his GPS signal intently for days on end. I did not know if he had been able to see the storm coming. Toward the end of the wilderness, we were able to talk randomly -- some of the higher peaks actually had a cell signal.

We needed to meet up with him in Millinocket, Maine. Hello Kitty, Trail Momma, and I piled into the car and headed off on the 1,200 mile journey. We were relieved when we got the call from Indiana Jones that he had made it there safely as we were traveling to Millinocket. The day that we brought the family together was amazing. I was so relieved that he made the rest of the trail on his own, and everyone was glad to be reunited after five months. When we met up, we hugged and I could tell from the look on his face that he was glad to see me, and I knew the last week or so was probably very tough on him with all the rain. He had 10 miles left to hike from Abol Bridge Campground to the Birches and then the final 5 miles up to Katahdin and back.

I hiked with Indiana Jones that first 10 mile stretch. I didn't know whether I could even do it because the foot was still very tender. The last portion of the trail was the ascent of Katahdin – this would be no small task, a 4,000 foot vertical climb, steep rocky terrain, and most of the elevation gain happens in a mere three or four miles.

I waited until the day before the final climb to even make the decision to try it. The foot was just not healed that well, but I finally decided to go. How could I not? This was the pinnacle of the trip, and even though I didn't hike the entire AT, there would have been a lack of closure if I had not tried it. The final day we started at 9:00 a.m. Hello Kitty, Indiana Jones, and I made the climb together. We also found out that our Swiss friends, Chocolate and Cheese, went up that day too.

When we got to the top it was awesome! We celebrated with many other hikers -- what a feeling of excitement! I could tell Indiana Jones was really enjoying the moment. He did the whole thing, every single step of it! As for me I am proud I did as much as I did. I hiked over 1,800 miles total which is pretty decent. I went back in 2015 and did another 100 miles but ended up with several injuries again and had to once again stop the hike.

This was an awesome journey. There were emotional ups and downs during the last part of the trail, but I wouldn't trade it for anything. Doing it with my son was an incredible, once in a lifetime opportunity. After the trip, Indiana Jones and I were asked to speak at many engagements about our journey so we could relive the trip a few more times. Amazingly, people would listen to us with keen interest even if we talked for an hour or more.

Today, two years later, I don't think there is a day that goes by that I am not reminded about something we did on the trail and how detailed my memory is of all the places we visited and the people we met. Many have asked whether I would do it again. My answer is generally "no", but only because I don't think my body could physically take it! If that wasn't an issue my answer would be "Hell yes!"

Some additional thoughts from Indiana Jones

Hey all! I hope you had a great time reading the chapter that Big Dawg wrote. I went through his chapter and added my own edits and elaborations, but he did a great job of capturing the essence of our journey. Not all of the details about the people, places, and things that we saw made it into the chapter, but we didn't forget about them. If anyone reading this was someone we met on the trail and was not included, don't worry! We still remember you.

I want to thank Big Dawg for everything that he has done for me – without his support and preparation, our adventure would not have happened and been the same memorable and unique experience. I wish that we could have done things differently so that we could have both finished the trail and taken some more time, but as he said – I wouldn't give it up for anything else. Thanks again Big Dawg – you are a great hiking partner and a great father!

I will add a couple of thoughts here about the time I spent hiking alone. After Big Dawg got of the trail the first time, it was a very strange experience for me. I had gotten used to hiking with him and working as a team to set up camp and the talks we would have at camp and during our breaks. His absence was very noticeable, but I never felt alone. I was always hiking around other thru hikers that I knew and I stayed at shelters with people that I knew. Even with him not on the trail with me, the hiking did not change too much for me. Sure, I could hike a little bit faster and longer and make some different decisions, but overall it was still pretty much the same as hiking with him.

For more details about our adventure, you can search trailjournals.com for Indiana Jones and look for the entries from 2014. Additionally, the link is listed below:

http://www.trailjournals.com/entry.cfm?trailname=16807

~ ~ ~

Al Olsavsky aka "Big Dawg" is retired from Delphi Automotive. He is an engineer by degree and has held many engineering positions in his career. Later in his career he transitioned into the plant manufacturing operations. In his last role before retiring he was an Operations Manager at an electronics plant in Kokomo, IN. His introduction to the Appalachian Trail was as a Boy Scout in the Washington DC area. He completed many hikes on sections of the trail as a Boy Scout and had thoughts even then about some day doing the whole trail. He was an Eagle Scout. He rejoined the Boy Scout movement with his son, Shawn, in 2002. He participated in many outdoor camping/hiking trips and other "high adventure" activities. During that time Al and Shawn talked about maybe doing the trail someday.

Shawn Olsavsky, aka "Indiana Jones", is a recent graduate of Purdue University. His undergraduate degree was in Aeronautical and Astronautical Engineering. He went on to earn a Masters degree in engineering management from Purdue. Shawn has always been an adventurous outdoors person. He completed many activities with his Dad as mentioned above including a 10 day 65 mile hike at the National Boy Scout Reservation known as Philmont. Since completing the trail in 2014 Shawn has earned his graduate degree and now works as a test engineer at ATA in San Diego, CA.

Chapter 13 **Blood (Mountain Cabins), Sweat and Tears**
Kate Spillane Balano aka Katwalk

Why?

When people ask me why I wanted to hike the
Appalachian Trail – all of it, in one go – I didn't have a
good pithy answer. I blather on about this and that.
Because, the reality is not simple and the seed was
planted years before I began the hike.

Approximately 90 miles of the roughly 2,190 miles of the
Appalachian Trail meander through my home state of
Massachusetts. Crossing rivers, streams, and postcard-
picture towns, the trail climbs the highest peak in the state
(Mt. Greylock) and is frequently considered a "welcome
back to real mountains" wake-up call for thru hikers
approaching northern New England. This part of the state
is the summer home of the Boston Pops and the wintry
Massachusetts of *Ethan Frome* and Sweet *Baby James*.
Yet, while only 50 miles distant, these mountains are
light-years removed from the industrial city of my youth.

Every fall an odd sort of pilgrimage begins in New
England. People from all over the US, as well as native
New Englanders, hop into the family car to "see the
foliage". Mother Nature is arguably at her best in this part
of the country every autumn. The sugar maple and other
deciduous trees explode into reds and yellows and
oranges of incredible intensity. For us, this annual road
trip meant following the old Mohawk Trail. Originally a
Native American trade route, the Mohawk Trail is now
mostly along Route 2 and complete with shops selling
souvenirs of all kinds but with a big emphasis on maple
syrup and super-sweet maple sugar candy shaped like
maple leafs.

Between the towns of Williamstown and North Adams, just south of the Vermont border, these two trails intersect. As cars drive along Route 2, they climb upward, so that the beauty of the foliage is seen along the roadsides and, in some places, vistas of hill after hill after hill echo across the landscape. You might say that I always knew about the Appalachian Trail. It was off to the west in an area of amazing beauty. As a bookish kid reading *Heidi*, it was my only point of reference for the descriptions of the Alps. In my mind, the mountains were places of beauty complete with mountain streams, waterfalls of icy cold water, and fields of sweet smelling wildflowers.

It seemed the mountains were always calling to me – even though I didn't especially like to spend time out of doors. I cannot tell you how many times I watched The Sound of Music; I even walked down the aisle to the Processional from that movie. The only (adopted) child of older (and overprotective) parents, I was always more likely to be reading a book somewhere than playing outside. My only experiences with camping were both miserable. The neighborhood girls "slept over" at a friend's house in a tent smelling like their family dog. Rocky was part boxer and part who knows what, but one thing is certain, he had never been anywhere near a bath. And then there was Girl Scout camp -- sleeping in cabins with uncomfortable mattresses with no chance to read at bedtime. Being the nerdy, unathletic kid was not a big plus. Coupled with seasonal allergies, these were experiences I vowed never to repeat as an adult.

And yet, when I read of backpacking and sleeping in a tent along a stream I felt a surprising pull within. For some strange reason, it was something I wanted to do. I loved to walk. When I was old enough to work, I walked miles every day to get there. I walked to visit friends after we moved further from the city. I sneezed like crazy, but I still loved the freedom of walking.

Fast forward past the years of college and raising a family. Somehow I had become middle aged and beyond. My work schedule had made my walking sessions less frequent, and I had the knees to prove it. We moved to Florida, which meant that hills were not a factor in any walking that I did. Climbing stairs hurt; so I was glad to be all on one story. Stepping over the dog gate hurt. I started using an elevated toilet seat to make getting up and down easier. I went to a doctor to see if I needed surgery. Her advice was to move more. Surgery was inevitable, but exercise and physical therapy would help tremendously. I gave it a try, but the PT seemed to do nothing and my exercise efforts faded at warp speed.

In the fall of 2012 we learned that we would become grandparents – times two. During the weeks of helping my daughter, I struggled with going up and down stairs multiple times in a day. I realized that, if I was to be a fun grandma, some changes were needed. I started with knee strengthening exercises at home and with walking. Soon, I was up to four miles a day. And thoughts from somewhere, deeply hidden in my brain, began to drift into consciousness, whispering "Appalachian Trail" over and over like a mantra.

Maybe this Could Be Real

The Internet is a beautiful thing making it so easy to locate information on the trail. Eventually, I found the Trail Journals site; I had hit the mother lode! It seemed there were many men and women "of a certain age" attempting thru hikes. Their journals provided encouragement and sustenance. I read books recommended by other hikers and read installments in several journals daily. By the Fall of 2013, I felt bereft that the "AT season" was coming to an end. I was addicted and knew that it was time for me to take this dream to the next level.

So, like any good nerd, I created a plan:

Step 1 – Confront the husband that had heard me say many times that "my idea of camping was a hotel without room service." It is to his credit that he showed surprise, but never laughed out loud. Not even once. Those of you who have read my journal know him as St. James – a fitting title if ever there was one. I am also grateful that neither of my kids told me that I was losing it in a big way. They may have told others – "I'm thinking Mum is going crazy" but they never let me think I had anything but their support.

Step 2 – Go back to my doctor. I went to get my knees re-checked and (gulp!) told the doc why. She didn't laugh either. She was in the process of relocating to Atlanta and was excited to learn that the AT was so close. I told her my plan – to hike the Approach Trail from Amicalola Falls to Springer Mountain. She told me to have some ice handy at the end, just in case. She was the first of many people to tell me that I was amazing.

Step 3 – On December 23, 2013 St. James dropped me off at the top of the falls. I had decided to forego the 600 steps that start the approach trail, as I wanted to give my knees an idea of what the terrain would be like on the "real" AT. It was a misty gray day, and for the first few hours, I met no one. I sung along with Christmas music on my iPod, until I spooked a deer and a couple hiking southbound. It took much of the day to reach the summit – partially because I had promised to stop every hour to let St. James know I was okay. The only difficulty I had was in finding the parking lot off the Forest Service Road. Fog was rolling in and I couldn't figure out where the lot was. Somehow I didn't realize that I had to hike along the AT to get there. Fortunately, a family came along and pointed me in the right direction. Note to self: buy the Guthook guides for all sections hiked and be sure you are aware of where you are hiking!!!

It was magical. It was wonderful. I didn't have swollen knees at the end – though I did drink a Vitamin Water and take some preventative "Vitamin I". On Christmas Eve I was walking like an old lady, but I remained elated. The plan was to join a gym in the new year and do regular stair climbing and weight bearing exercises.

So now the real work started. I started working out 5-6 days a week and researching gear when I could. I made lists and lists of lists. During the summer of 2015 St. James and I did some hiking in Maine. This resulted in the first real hiccup. We went to the library and found a trail in the Camden Hills near where we were staying. It was listed as a "moderate" hike of about 2.5 miles on a blue blazed trail. With no poles or boots, we thought it might be just what we needed. It was HARD! There were places we had to scramble up rocks and hang onto tree branches. And no matter how far we went, it seemed there were markers stating that the summit was a mile away. After a few hours of this, I was beginning to hate those Tiffany-blue blazes.

St. James took a tumble and encouraged me to go on to the summit (it HAD to be less than a mile away by now). The photos from the top were outstanding. Camden harbor lay before me, with tall ships at anchor. And FOG rolling in! Yikes. I turned around quickly and headed back to St. James. We slipped and scrambled our way down, making it to the car using our phones for light. Whew!

We decided that burgers and beer were in order, but not necessarily in that order. After finding a sufficiently dive-y looking bar, we did a bit more research on our phones. What we found is that there were multiple light-blue blazed trails going up that mountain. Some were considered to be difficult. Ok, I could continue my love affair with Tiffany blue. And, while that made me feel better, I knew we had to try another test – the AT itself.

Per the ATC website: "Maine is the AT's most challenging, rugged and remote state, and it has the wildest feel of any area of the Trail. Maine features some exciting features that are rare elsewhere on the AT, including wildlife like moose and loons and pristine lakes. It's also famous for hosting the hardest mile of the Trail: Mahoosuc Notch."

I wanted to find a challenging day hike (in and out to get the hills in both directions), but not so hard as to make me give up before getting started. I also didn't want to spend all day driving and needed an easy place to park. So, back to the library the next morning. It was a wet day – perfect for research!

After looking for hikes that met my criteria, I finally settled on Little Bigelow. Maine Trail Finder described it as: "This relatively short and accessible section of the Appalachian Trail affords users some wonderful views of the Bigelow Range and eastern end of Flagstaff Lake. A lean-to and nearby series of pools also provide interest along the trail." It was rated as Moderate/Advanced and a 6.2 mile round trip. The more difficult trails were rated Strenuous/Advanced, so this seemed a good test.

St. James and I had a blast. We bought some beverages and Snickers bars to do a little trail magic and I met my first thru hikers. They were all so encouraging and they made St. James feel more comfortable about my adventure. The bottom line was: Go for it! The race is on…

So, in September of 2014 I ratcheted the level of preparation up a notch – and welcomed a third granddaughter. She and her twin sisters were powerful motivation for me. As I put it in a blog article on *Appalachian Trials*, "…one of my reasons for hiking is for them to remember me as the most badass grandma ever."

With just over 6 months to hike, I had a lot to do. I really hate prepared food and it makes me break out in hives if I eat too much. Using everyday recipes, those I found on the Internet and those in the book *Backpacking Gourmet* by Linda Frederick Yaffe, I began to dehydrate and freeze meals. I bought individual packets of salt, pepper, and olive oil to add as I rehydrated them. I knew that when I was flat out tired, the last thing I would want to do was to spend a lot of time cooking food on a hiking stove. I found lots of tips on "cooking" in a freezer bag inside of an insulating cozy. I would not regret that decision.

On Trail Journals, I found two other hikers near to my age from Florida who were planning to start on the same day. One hiker, "Trip'nRoll," had to postpone her hike for a year. The other was BonBon. We talked and texted a lot in early 2015. BonBon is one of the most amazing women I have ever met and I was fantastically lucky to begin the trail with her.

St. James and I drove to Atlanta and then caught a plane north for the wedding of a dear friend's son. The groom was an eagle scout and that seemed a good omen. We flew back to Atlanta and I dyed a purple streak in my hair – what a mess! But it was a symbolic action separating the years of corporate America from my new year of adventure.

And So We Begin

We met BonBon the next morning and off we went. After a few stops – including Trader Joe's to get some of my fave treats and a week's supply of "Two Buck Chuck", we were off. We stopped at Amicalola to register; I was hiker 674. Then we drove to our home for the next few nights, Blood Mountain Cabins. One thing for certain – it takes forever to get anywhere by road in the North Georgia Mountains.

Katwalk and BonBon on the trail

When Trip'nRoll and I were planning our hike, her husband had planned to stay at the Blood Mountain Cabins and do some early slack packing for us. When she had to cancel, the thought remained in my head. I wanted to start in March to give me a timeframe to finish by early October, yet I was concerned about the cold weather for the first month or so. I checked with the cabins and they would rent us a cabin at a weekly discounted rate and put us in touch with a shuttle driver. All this cost more than we had planned, but it would provide the foundation to make our hike a success.

Every morning our shuttle driver, Mr. Sam Duke, picked us up at the Cabins' General Store. While driving around sharp corners and hairpin turns, he regaled us with stories from his Louisiana childhood and his life after moving to Georgia. We kept our packs about half to two-thirds full - "half packing". It gave our bodies time to adjust to eight-ish mile days up and down mountains before we needed to take on full packs. It was a decision that I never regretted. Having a shower each night as well as dinner at a table and chairs is priceless. Not to mention the Two Buck Chuck.

It didn't take long to discover that we were both slow hikers and that I was the slower one (by far). BonBon was so much stronger than me – and so mentally tough. Even with all of my preparation, I was still not fit enough to easily conquer those mountains. I knew in my brain that my body would improve its level of fitness the further I hiked, but, in my heart I was down. I had developed some minor blisters and they added to the feeling of discouragement. I was transported back to the geeky kid, who was always the last to finish any race. BonBon was my salvation! I know she waited patiently for me at the lunch stop and at the end of the day. More than once, she walked back to check up on me. Her sense of humor was as silly as mine. She once said that we get along so well since we both have the sense of humor of a fourteen year old boy! Being able to treat my blisters nightly and keep them clean (shower!!!!!) also did wonders for my feet and my morale.

A funny thing happened that week. Something I never anticipated. I lost my appetite. Trust me – this is not a woman who willingly misses meals. When I have a tough time eating Nutella and drinking wine, something is really off! Later I learned that a lot of hikers went through this. For me the solution was a small breakfast, mid-morning snack, small lunch, mid-afternoon snack, and then dinner. This kept me going and made walking easier. With a full stomach I always felt it was harder to breathe on the uphills.

The Blood Mountain Cabins turned out to be a pretty social place. We met many hikers that we "knew" from Trail Journals. One of them, No Spring Chicken, a recently retired nurse and I had talked prior to the start of the hike and she stayed with us a couple of nights and shared our shuttle. About six nights in the weather turned really, really cold. We decided to take a zero and were so thankful for our little cabin. The temps dropped into the low teens overnight. The heat in our cabin wasn't working, so the owners gave us two space heaters that really did the trick. They were willing to move us, but there were so many hikers trying to find a warm space that we said we were happy with the space heaters. That night a group of us pooled our resources and cooked a spaghetti dinner together. It was an absolute blast. It would turn out that social aspects of the hike were a huge factor for me.

On Our Own

We got a bit teary when we said farewell to Sam at Tesnatee Gap. It would be a short day hiking to Low Gap. Rain was expected that night – not what I wanted for a first time using my tent. I was so pokey stopping to look at the beauty of the woods that there was no room at the shelter. We set up our tents and cooked our first meal in the woods. I messed up big time with my tent. I had the stakes positioned incorrectly and I got wet overnight. The storm that came through sounded like a freight train through the gap and was accompanied by thunder and lightning which struck near the shelter. BonBon had me

get into her bag to warm up and made me some hot
oatmeal. It tasted terrible but it did the trick. Not wanting
to risk a wet bag that night and knowing it wouldn't have
time to dry, I had Sam pick me up and take me to the
Holiday inn Express in Hiawassee. I was able to wash off
and dry my stuff.

BonBon stayed out and camped all alone that night – she
really is a warrior woman. We met up the next day when I
slack packed southbound to make up some distance. After
taking a zero in Hiawassee we set out together again and
shortly we were done with Georgia! On our first night in
North Carolina I got wet again. This time I knew the
water was coming in through the sealed (!) seams of the
tent. I was able to get my stuff pretty dry at lunch, but the
night was going to be very cold again. I got nervous about
staying warm enough. My bag was rated to 15 degrees
and I had a liner. BonBon's bag was rated to a higher
temp and she didn't have a liner. I knew she was awake
and asked if we could room together that night. It was a
wise decision. I think we would have been very cold
apart. We didn't get tons of sleep, but we stayed warm.

Breaking Up is Hard to Do

Another blog that I wrote for Appalachian Trials prior to my hike was all about hiking partners. It seemed to me then, and still does, that getting a good hiking partner was much like dating (without the sex). You needed to find someone that you were comfortable with. That person needed to hike at about your pace. You needed to be in sync about tenting vs. shelters. Ideally you would also have the same philosophy about slack packing. You also needed to agree about finances.

BonBon and I continued to get stronger and more confident. In many ways we were prefect hiking pals. But if I followed my own advice, there were a few areas where we were out of sync. I had no issues with going southbound to slack pack. Many times this wasn't to avoid a tough part of the trail; it was simply to get to/from a hostel by the most efficient way. BonBon wanted to be a pure NOBO. We respected each other's decisions, but it was a small difference. As mentioned, while she thought herself to be fairly slow, she was considerably faster than me. In the early days, I was very cautious on the "downs" until my knees got stronger. She also wanted to finish the thru hike much sooner than I knew I was capable of. She had a daughter starting a new school and a business to run.

I was afraid that I was being very selfish; that my slower pace might negatively impact her timetable. I was also very concerned about the Smokies. I hated the cold and wet. Prior to starting our hikes, No Spring Chicken (NSC) had offered me the opportunity to skip ahead and then come back and finish the Smokies in early May. Meanwhile, BonBon made the decision to stay at the Nantahala Outdoor Center. Up until then, I had avoided crowded bunkhouses and shelters due to reports of the Norovirus early in the year. I have Crohn's disease and

Noro could have ended more than just my hike. I accepted NSC's offer and regretted that decision almost as soon as I made it. I missed hiking with BonBon terribly. It was really a horrible "break up" – even if I felt that I was doing the right thing for us both.

New Beginnings; Tough Decisions

NSC was fun to hike with, but BonBon and I were sisters of the heart and soul (sole?). NSC and I skipped from Fontana to Hot Springs thanks to shuttles and accommodations from her husband and sisters. Hot Springs was a great town and we enjoyed some good food (i.e. burgers, pizza, and beer). We were able to slack pack a couple of days and enjoy the break. We then hiked to the Black Bear Resort, near the Virginia border.

In early May, NSC's brother flew in from Washington to hike the Smokies with us. He was a strong hiker and had hiked this part of the AT while he was in Med School. The weather in the Smokies was absolutely wonderful. I later learned how awful it was when BonBon went through and I felt horrifically guilty. NSC found the pace to be difficult. We were able to stretch out our food and extend our time there by a day.

As we began to descend toward Hot Springs it began to get much warmer. My feet had been hurting and NSC showed me a type of shoe insert she used. I tried them and they immediately relieved the pressure on the balls of my feet. Without them, I'm not sure I would have made it that far.

Just prior to Hot Springs, NSC decided that she would get off the trail. I was so sad. I hiked solo into town and spent one last night with her and her brother. I was really beginning to doubt my own desire and ability to finish.

My feet really hurt and there was a lump on the arch of my right foot.

The next day I moved to the Hot Springs Resort as my daughter and youngest grandbaby were coming to visit. The day they came it poured down rain. I was so glad to be taking a zero. I probably cried at least as many tears. However, I think my decision was (at least partially) the right one; head back to where I left off and try it on my own.

At Black Bear I ran into a neighbor of mine (Grandbob) who had tried to thru hike a few years earlier. He was attempting to resume his hike, but couldn't feel the vibe. He never got to know any of the other hikers as they had their "hiker legs" and he was just getting started. He was leaving in a couple of days and offered me a ride if I wanted one. I decided to hike one more day and see how I felt. It was one of the most beautiful hikes on the trail. It was also in an area of high bear activity, so I was hiking SOBO from Wilbur Dam back to Black Bear. My foot did not do well. I kept stopping to adjust the insole to try to minimize the pain in my arch and ball of foot. I hit a spot where I wasn't sure where the trail went – and burst into tears. It took me forever to walk those miles and, at the end, I was pretty sure it was time to go home. I ate a pizza in my cabin and was totally antisocial that evening. I kept second, third, and fourth-guessing my decision. But reality was, there was nothing I could do about a lump in my foot and horrible ball of foot pain except getting it checked out or hike through it. I was concerned that, if I kept hiking, that I might do some permanent damage. I didn't want to be a thru hiker at 60 and in a wheel chair at 70. Yet, I still felt like a failure.

Taking a Rest

I went to visit friends with Grandbob and eventually got home to get my foot checked. The podiatrist gave me a cortisone shot for the ball of foot pain and suggested surgery for the lump. Surgery was, in my opinion, pretty much out of the question. He talked about a limited chance of success and disconnecting the *Plantar Fascia* – it would take a lot more pain to make me go through that. I missed the trail as if a part of me was left behind. I couldn't settle into anything at home. I spent some time with my little ladies and thought constantly of the trail. In a short time the lump had receded to a very small size. I was ready (I thought) to try again.

Take Two

I proposed to Grandbob that we hike together. This would give him the companionship he wanted and it seemed like we would click on the trail. We headed up to Pearisburg, Virginia, where he had left off. If we made it all the way to Maine, I would need to come back and finish the section from Wilbur Dam to Pearisburg. It was wretchedly hot and there was little water, but I was so glad to be back. I quickly understood why Grandbob had felt a bit lonely earlier that year; all the hikers were doing 20 mile days while we were doing 12. Grandbob was more of a shelter guy, and since the Noro risk was much less this time of year and the shelters were also less crowded, I slept in them too. One of the things that I had learned with my Crohn's is that life on the trail is easier if you take time in the morning for a few cups of tea. Being near a privy helped, but the best was not having to stop and take my pack off repeatedly all day long.

Grandbob was feeling the heat more than me, but we were otherwise doing well…until one afternoon, a few days after we started, when his knee popped. He knew he had to stop hiking. He offered to take me further north to connect back with BonBon. I knew she was in the Shenandoah and would shortly be taking some time off to spend with family. I also knew she was putting up bigger miles than my feet could handle. So I decided to strike out on my own.

Sometimes I think I am really, really dense. I hated hiking alone the last time – why did I think it would be different this time? The young hikers I met each evening continued to disappear the next day. I kept going, but my heart wasn't in it. My foot hurt somewhat, but I think the loneliness made it seem worse that it was.

Home Again, Home Again

This time I felt less ambivalent about going home. But soon after I got there and I analyzed again and again what went wrong, it all came back to needing that hiker comradery. I hated being alone, nearly all day, every day. Gradually, I came to realize that I should have joined back up with BonBon. Even the parts that hurt are mitigated by having such a good time. So what if I was last in to camp every night? I followed her progress through Trail Journals and did all I could to support her, especially by sharing dehydrated meals that I still had. She was running low on energy, and I kept thinking part of that was due to trying to get by without enough protein at dinner.

I started picking up the threads of my life again. I had hiked over 500 miles on the trail, but deep down still felt like a bit of a failure. It didn't get any better when a neighbor commented, after seeing *A Walk in the Woods*, that Redford and Nolte's characters were a bit too old for that type of adventure. Then she looked at me and said, "After all, it didn't work out so well for you either, did it?" A friend spoke up right away and protested that what

I did was incredible, but still the comment stung, feeding the feeling of being not quite good enough.

The Maine Finish

BonBon had been asking if I would come to Maine and be with her at the end. D-No and C-Shell, who she had hiked a thousand miles with, had just gotten off the trail due to injuries. BonBon and I texted at night a lot and I was with her in spirit. Her enthusiasm was so infectious that I made the decision to join her in Maine and hike at least a part of the 100 Mile Wilderness with her. However, I left climbing Katahdin to be a game day decision.

I flew to Portland and was picked up by my sister-in-law who lives nearby. We had a good, brief visit and then I got picked up by a shuttle driver to take me to Stratton. BonBon joined me the next day and we had a great evening at The White Wolf Café in Stratton, enjoying our fave trail meal – burgers and beer. The weather was lousy and we put off starting for a bit. Somehow I had gotten my geography messed up as I wanted to start after the Bigelows. I decided to zero rather than start off on a really tough hike.

We met up again in Stratton and it was evident that, even though I had been doing some workouts, I was not anywhere near as well conditioned as BonBon. When we left Stratton, I started a ways up so that she could catch up to me. Because of future plans, she needed to be home the first week in October and I didn't want to hold her up. I ended up in Millinocket first (mind you, only because I skipped about 40 miles ahead). When BonBon got there, she really wanted a rest day before Katahdin, but the weather later in the week was looking threatening. So off we set on September 28.

The hike through Baxter may have been the easiest two miles on the trail. And then the fun began. Up, up, up we went over rocks and dodging a stream running down the trail. We had planned to get a ride from a trail angel, but that got cancelled. It would have given us plenty of time to summit and return. Because we were limited by shuttles, I stopped near the Tablelands and took a lunch break and then started down. BonBon continued on and summited with many of the hikers she had hiked with for many miles.

Not waiting a day turned out to be fortuitous. The rain poured so hard that a state of emergency was declared in parts of Maine. Hikers were stuck between the fast rising east and west branches of the Piscataquis River. We were forced to take several detours on the ride to Bangor.

BonBon and I spent our last day in Maine watching *A Walk in the Woods*. We roared with laughter at the idea of someone packing a baby blue robe and getting a lift off the trail in a golf cart. It was a great ending to an epic adventure.

Perspectives Eight Months On

As I write this, I realize that I have learned a great deal from my AT experience. Even being slow, I was in better shape than I had been for a long time. I didn't want to lose this. I had lost about 15 pounds on my hike and gained about five back. I began to do a Boot Camp routine and felt myself getting stronger. I kept wishing that I could find a cardio workout that I liked as much as boot camp. From my daughter, I heard about a fitness routine called Orange Theory that is a type of interval-based cardio and weight training that focuses on your heart rate as you work out. A year ago, I would have been too intimidated to try, but I tried it and loved it. I was amazed how well I kept up with much younger people.

I also began doing yoga and I could see my body begin to change its shape. I had definition in my arms for the first time ever. This motivated me to begin to eat healthier. My daughter began working with a nutritionist and I started as well.

Last week BonBon, C-Shell, and I did a four day hike in the Georgia Mountains. I was amazed at my energy level. I wasn't sore. I wasn't tired and I flew up and down mountains that were difficult last year.

This has me thinking. If I hiked 600 miles at sixty, maybe I can do the rest by 70?

~ ~ ~

Kate Spillane Balano – aka Katwalk – hopes to be a New York Times best-selling author with the publication of this book. She has worn many hats over her 61 years including accountant, mom, professor, sales director, wife, grandma, blogger, and hopeful AT thru-hiker. Prior to that first step from Springer Mountain, her long-distance hiking and backpacking experience was nonexistent. Next summer, Katwalk plans to hike the Coast-to-Coast Walk across Northern England.

Chapter 14 Funnybone Comes Back
Jim Dashiell MD aka Funnybone

There are many Appalachian Trail experiences, as many
as there are books about them. Whoever writes the next
story can be certain that it will be unique to them. The
author may be the only one who agrees with the whole
story even after memory has modified some of the events
or people, since time fades color, bends the truth, and
redirects destiny. That is how it should be. We all have
our own life stories to tell, our own realities. The
explosion of events, of moments on the trail, leads each
hiker to different conclusions just as each summit has its
own views, its own history. The trail experience will be
many things to each hiker. It is almost mystical how
events happen on the trail, beyond serendipity. It can be
punishing and unforgiving to some, a photographer's
utopia, a palpable calm to the troubled mind. It is also an
environmentalist's playground, a naturalist's Nirvana, an
historian's living museum.

All hikers start as equals. Life on the AT is like living in
the Golden Rule. Society's greatest sins, greed and the
quest for power, are missing on the AT. The slogan used
most often is "hike your own hike". It is an expression of
individual freedom for each hiker, respect for their
decisions and tolerance for their individuality. If all
members of Congress were forced to do a thru hike on
their own before they could run for reelection this country
might regain the passion and wisdom of our founding
fathers.

The trail is free from the curse of boredom, which, like
rats, is endemic in the cities. Individuality is a prized
characteristic of the hiker, sometimes to the extreme.
There is a camaraderie that develops among hikers out of
respect for shared hardships and rewards. It is like a club
without charter or bylaws.

My inspiration for hiking the Appalachian Trail grew slowly, almost without me noticing. I read an excellent book by Jean Deeds, *There are Mountains to Climb*, that describes her hike of the Appalachian Trail in the mid 90's. She made it sound like a difficult, rewarding but doable life changing event that might be within my abilities. Jean is one of the contributing authors of this book!

I had retired at an early age (53) from an orthopedic surgery career. I was married to LaVonna, a nurse who was adventuresome, loved to travel, but was frugal and content with "home" things. Our blended families lived close. I have 10 grandkids, all living within a 45 minute drive. Life was very good. Then LaVonna became ill, was diagnosed as having malignant melanoma with brain metastases and died in less than five months. I took care of her to the end with good help from her sons and their families. I carried some of her ashes with me from the beginning of my hike to the top of Mt. Katahdin where I spread them along with tearful prayers. I was holding her the night she looked around our room and asked, "Who are all these people?" The room was empty to my eyes. She smiled a big smile, understanding what was happening, and breathed no more. Her death was simultaneously the 'end' for me but also the 'beginning'.

After several years of a busy but directionless life the challenge of hiking the Appalachian Trail caught my attention, but only a little. I didn't really know what it was, so I read several more books after Jean's had planted the seed. The more I read the more I was drawn to the prospect of leaving my comfortable, predictable life and taking the leap, so to speak. I had only backpacked for a total of three days and never even slept in my tent until the first night on the trail!

The call of the AT, the adventure, the unknown, the "bucket list" type of challenge, the thought that my 10 grandchildren might someday appreciate what I had done (their parents, too), grew in me until I was debating with myself the real possibility of making the commitment. It wasn't until I had actually visited the Trail as it passed Clingmans Dome in the Smoky Mountain National Park, walked along it for 100 yards or so, sat down beside it and said a prayer asking for guidance, that I felt not only that I could hike it but should.

While I do consider myself a Christian I had not felt the presence of God before in such a palpable way. I was encouraged to attempt the hike. I kept those thoughts to myself for a couple of weeks, would not think about it, blocked it from my mind. Then I sent an email to all my kids, their spouses and my oldest two grandchildren, Paige and Nate, stating my plan. Several days passed with no comments except from Paige. She had just turned 13. Her one-word answer was "AWESOME". When I asked the parents later why none of them replied I was told they were trying to decide at whose house to hold the intervention. And so the decision was made.

While March 26 of 2012 was the day I walked away from the southern terminus of the AT on Springer Mountain, my preparation for the adventure began about a year earlier. I started to explain to friends what I had decided. Their questions, encouragement, and general confusion about just what this meant brought back to me the enormity of my decision. It forced me to face the reality that I might fail, might be ridiculed for thinking such an effort was possible for someone who had never backpacked at all, who had only slept in a tent once while I was a Boy Scout about 55 years earlier. It grew from a silly notion into a passion. Why did I want to do this? Was it merely some attempt at self-promotion? Possibly an ego thing? Such negative thoughts were not without merit. A nurse friend, Carolann Shaddy, gave me my trail name. I had been a "bone doctor", and I guess she thought I was funny. So, Funnybone it was!

The fact that only a small percentage of skilled backpackers are able to complete a thru hike made it an extreme challenge for me. I enjoy taking the difficult path, the road less traveled. I can do this, I told myself. Jean's book and Jean herself were supportive and encouraging. I felt that with good luck and the right attitude a successful thru hike was possible. Boy, was I right about luck and attitude! Many excellent hikers I met had to leave the trail because of injury or illness. Some just became burned out, as I did briefly in late August. I quit the trail near Gorham, New Hampshire, went home for a week and a half, got my mind right, replenished my energy reserves, and returned to the trail. I summited Mt. Katahdin and then headed south to Wildcat Mt. I'll say more about that later.

I found that I liked to write in my tent in the evening or during a rainstorm, so I'll sprinkle a few poems throughout my chapter. I'll start with a couple of limericks.

Life on the Trail

Life on the Trail can be cruel
No water, no food, no fuel
Solitude all day long
Surely something is wrong
Who wants to live like a mule?

Then comes a blessed moment
The calm, the peace, you own it!
When your thinking turns good
And you know that you would

Receive the trail's atonement.

Not only was I able to complete my goal of a thru hike by doing a flip-flop, which I prefer to call a BiB0 (goes both ways), and heading SOBO, it allowed me to say a proper goodbye to about 60 thru hikers I ran into who were continuing NOBO. These were many of the bubble I had hiked with, met in towns or shelters, and generally was familiar with, some from my very first week on the trail. That was an added bonus I had not expected when I returned. Many said they had heard I was off the trail. Of course, they had heard correctly, but Funnybone was back! I was able to tell them what lay ahead, the fantastic climb up Mt. Katahdin, and to encourage them, even prematurely congratulate them on their awesome achievement.

I felt like a one man cheering section. It was a bitter sweet time for me as I knew I would likely not see most of my thru-hiker friends again. Our meetings were always a pleasant surprise. The conversation centered around other hikers in the vicinity, news of who had finished or gone home. I learned not to look back as we parted. That's not the last memory I wanted. It was sort of like when I go to a funeral. I don't care to see the deceased in the coffin. I prefer to remember them as they were in life. The older I get it seems I go to more funerals and fewer weddings.

One of the hikers I met as I headed south was Johnsie. He was one day from Katahdin. I shared a room with him in Franklin, Tennessee in early April and had not seen him since. Now, here he was five months later. He had attempted a thru hike the year before but ended up in the ICU of a hospital in Pennsylvania. He started over this year from the beginning and was practically in Mount Katahdin's shadow to complete his thru hike. Congratulations Johnsie!

I also encountered Turtle. I had hiked with him several days in the beginning and very much enjoyed his company. He too had attempted a thru hike several years earlier but was knocked off the trail with knee problems. I had not seen him in about four months, although we had emailed occasionally, so I knew he was progressing steadily. Our reunion was a definite blessing. Being from Wisconsin he changed his name to Cheesy Turtle when he heard there was another Turtle on the trail.

Another hiker I met repeatedly along the trail was Yukon, one of the contributors to this book. I met him at Neel's Gap the first time, then at Blueberry Patch Hostel, then Fontana, the Smokies, Franklin, Tennessee, outside Greenwood, New York, Upper Goose Pond cabin, Massachusetts, then in the Hundred Mile Wilderness at Shaws hostel. He was always smiling!

I think now is a good time to elaborate on why I'm headed south in the Hundred Mile Wilderness. As I progressed through Vermont and the White Mountain Range in New Hampshire, I was falling a lot. I was having trouble picking up my feet and was, instead, shuffling. Bad idea on the trail. I had scabs all over my arms and legs but had not really gotten badly hurt. I had lost about 50 pounds. There was this 'message' that frequently came into my mind saying, "You're going to go home in a cast or worse." That was annoying. When I struggled up Wildcat Mountain just past Pinkham Notch on a rainy day, I was at my lowest point. That message was screaming at me! Then I saw it. A skiing gondola that was giving hikers a free ride off the mountain!

I had learned to listen to my intuition, to believe in it, and to trust it. I said goodbye to Gipcgirl, my hiking partner at the time, rode down, hitched into Gorham with Phoenix Rising, a veteran hiker who gave me some good advice, caught a bus to Boston the next morning, then a plane to

Indianapolis, where some of my family was waiting. In just a little over 24 hours I was home! Amazing! After hiking 1,838 miles in 5 months I had quit.

But not for long. I got my mind right, rested, talked with friends, but could not shake the feeling of loss. It felt worse than death. About 10 days later I was headed back to the trail, this time to Mount Katahdin. After I resumed my hike by summiting Katahdin I headed south, officially becoming a BiBO, meaning I had hiked both ways. Meeting many of my bubble in the 100 Mile Wilderness, as mentioned above, was a bonus I had not expected. Ready for another poem?

Funnybone is Back!

Beat me up, sent me home
The trail thought it had won.
I rested, ate ice cream
But knew I was not done.

I got my mind right
Felt the consuming yearn,
Started making some plans
For my certain return.

They say the trail teaches
It will wait for you.
The lesson I learned
Was that I was not through.

My friends all praised me
My family was relieved
But a quitter I'm not
In my heart I believed.

"Would you think less of me"
I asked my closest friend?

"Just hike your own hike.
It's your choice. Be a man."

I sped back to the trail
Made Mt. Katahdin my own.
My wife's ashes, some prayers
I left on this throne.

A list of my trail angels
I'd been keeping since the start
I read to no one special
But it warmed my heart.

So from the depths of despair
To the summit of Katahdin
I was renewed, refocused
Avoided what could have been.

Second chances are precious
They provide for the best.
A purely human emotion
I've truly been blessed

The trail was beautiful with all of the color in the trees but in September it becomes very empty. The days were shorter, the weather colder and wetter. I was aware that it was also more dangerous. The old adage that, if you were injured on the trail, just wait and someone would come along, did not really apply now. The shorter days, dreary weather, isolation, and my desire to just get it done kept me grinding out the miles. That's hard to do in southern Maine! When I finally did return to Wildcat Mt. on October 6th, the gondola was not running! I had been so looking forward to a victorious ride down! However, I caught a ride with a guy who was getting off work and was driving his truck down the service road.

So ended my 6 ½ month adventure. I spent the night in Gorham, then met Bev and Linda, friends I had hiked with from Rhode Island, at a diner in western Rhode Island for a celebratory breakfast before heading home.

Before I was single and living alone I would never have been able to abandon my responsibilities and run off to the woods for six months. Still, I could not have made it without a great team to help me. Everyone back home I asked to help did so with kindness and generosity. I always appreciated them! If it takes a village to raise a child, it certainly takes a team to assist a thru hiker!

My daughter Julie Horoho received my mail, paid my bills, handled my banking, and sent me cash and other personal items. She even came with her husband, Tim, and kids, Sydni, Dylan, and Lizzie, to get me off the trail for the first time in early May around Boone, North Carolina. We stayed at a KOA, visited Grandfather Mountain and Linville Cave, drove on the Blue Ridge Parkway, and generally had a very relaxing time with plenty to eat. They dropped me off on Roan Mountain. That was a tough place to resume. My son, Marc Dashiell, mowed my yard, checked on the house, and cleaned up after storms had trashed my yard. He was also very good about checking up on me as I was hiking. He also sent me items as I requested them.

My stepson, Brian Arrowood, along with his wife, Michelle, and kids, Paige, Kyle, and Haley, picked me up in early June near Lexington, Virginia in his large RV! Talk about a sudden change of fortune! From hiking up and down muddy trails, sleeping in a tent, and drinking water from a creek, to being driven to town in a comfortable RV -- it was great! We stayed at a nice KOA where his brother, Derek Arrowood, Derek's wife, Kim, and kids, Nate and Ellie, showed up a short while later. Being surrounded by family I loved, eating town food, sleeping on a bed, taking a shower, and just relaxing was so restorative. We explored the Natural Bridge Park and Caverns and did the Natural Bridge safari, a drive through a large park populated with camels, bison, antelopes, and other wildlife. It was tough the next day when Derek and Kim dropped me off. I shed a few tears as I trudged away up the mountain.

Derek and family returned to get me off the trail in Boiling Springs, Pennsylvania in early July. We spent the weekend at Gettysburg, touring the Civil War battleground in a bus and then spent several hours in the awesome museum. This time there were no tears shed by me as I was feeling much better about the hike and the adventure in general. I had been on the trail almost 3 ½ months.

Neil Planalp, a local farmer who works my land, also mowed my pastures. Andrew Sims, a young man I know from church, checked on my house and used one of my gardens as his own with my blessing. My sister, Jane Bledsoe, and her husband, Garland, were happy to mail me items, such as food and clothing. My brothers John and Jeff checked on my property. Terri and Allen Stout were especially helpful. Terri had worked for me as my typist and x-ray tech for many years. She took care of my little Yorkie, Sophie, while I was gone. It sure was a relief to know Sophie was in good hands.

I had initially thought I would do a blog on *Trailjournals.com*, but lack of skill, battery power, and reception made this an impossibly difficult task for me. The solution was for me to call Terri and dictate each day's blog to her phone when I had reception. She would type my message from her phone and send it to *Trailjournals*. She knew that if I was calling, don't answer. Her husband Allen helped out with the typing when Terry became too busy. Allen even drove me to the start of my hike at Springer Mountain. He was my first trail angel. (He got to meet Scribbles and her husband Garry. Scribbles is also a contributor to this book. More later about her.)

I'm active in the Emanuel Lutheran Church in Tipton, Indiana. The church secretary Mary Heffelmier copied my daily blogs and put them in a large ring bound notebook in the narthex of my church. Any members of the congregation who were interested could read about my adventures and follow my hike. When I returned home, Mary gave me that folder so I would have a permanent remembrance. That is now one of my most cherished possessions. Thanks again, Mary!

My fishing buddies, Jim Elder and Jim Cromer M.D., were frequent sources of encouragement. Elder was my weatherman. I would text him the ZIP code of where I thought I was and he would send me the weather report, especially when the weather looked threatening. When you're outside all day, storms, especially with high winds and lightning, are something you need to be aware of and smart about. Hiking on the top of a mountain during a lightning storm carrying a metal rod in each hand is certainly risky, even foolish. More than once I laid my hiking poles down and took cover some distance away while the storm raged.
Here's another quickie.

The Golden Rule on the Trail

Why does the trail with it's rocks and roots
Seem like a refuge to me?
The ups, the downs, mud over my boots
Yet it's where I want to be.

As sure as can be, and I'm no fool,
Random kindness from strangers who care.
It feels like I'm living the Golden Rule
Undeserved but present everywhere.

I've said a prayer for all who've known

Me on my path to the end.
Protected, approved, guided and shown
These mercies as from a friend.

Is the Golden Rule a living thing?
Why isn't it my morning prayer?
From God or strangers I've learned to sing
My thanks for gifts found everywhere.

Trail Angels

The reality of strangers doing SO MANY random acts of
kindness to/for me was a concept I was not prepared for.
Yes, I had read about trail angels and their generosity in
many books about the AT. But I was still amazed at the
frequency, the sincerity, the extent of their compassion.
Some of these generous acts required considerable
preparation and expense. There was a couple from
Stevens Point, Wisconsin representing a church group
called Feast in the Forest. I believe their names were Mo
and Mike. They had a very elaborate, plentiful supply of
free food, including sausage (links or patties), eggs
cooked to order, fried potatoes, juices, cheeses, fruit,
snacks, and chairs. Yes, I said chairs. I had been on the
trail about a week and didn't realize how much I missed
sitting on something other than a log or the ground. They
also had a cold keg of beer! Remember, they were from
Wisconsin. The most common food supplied by trail
angels is hot dogs, canned drinks, donuts, chips, and fruit.
One Lutheran Church in Virginia set out lemonade,
apples, and hard-boiled eggs. Each egg had a verse of
scripture written on it as an apparent encouragement to
look it up in your Bible. I did not carry a Bible but did
have one on my Kindle.

I kept in my journal a list of the trail angels I met along
the way and read that list out loud as I stood on top of

Mount Katahdin. No one there listened to me or cared what I was doing, but I knew these trail angels were a critical part of my journey to the end. I felt I needed to honor them in some small personal way. Not all trail angels fed me. Food was a great form of what we hikers referred to as trail magic, but that term also includes picking up dirty, stinky hikers for a ride into town, providing advice or directions, giving us a place to stay and clean up, etc.

I bet I rode in the back of pickup trucks 10 times. There was something special about throwing your pack into the back of a truck, climbing in with other hikers, the driver's dog, and perhaps some bags of trash. I think I had such good luck getting rides because I was older than the average hiker by several decades. There weren't many rides in trucks once you got past Harpers Ferry.

One of my favorite trail angels, who typified the whole phenomenon, is Stephanie of Boiling Springs, Pennsylvania. She is an attractive lady who collected me and Groove and Bonobo, a couple from Portland, Oregon, off of the trail. Stephanie let the three of us stay in her house, clean up, do laundry, and then visited with us that evening. She was a gracious hostess. Two of her children that still live at home were not the least surprised when they found strangers in their house. I asked her daughter if Stephanie often invited hikers into her home. The answer was yes. In the morning, while Bonobo and Groove slept in, Stephanie fixed excellent blueberry pancakes for the two of us and served them after grace was said. She lives the Christian life. By the time I met her I had been on the trail over three months. She told me, accurately, that I would look younger if I trimmed the beard. I loved her honesty.

At 66 I was the oldest thru hiker I knew of in 2012. There were four guys that were one year younger, G0 man, Missouri Mule, Nooga, and Johnsie.

I ran into a unique form of trail magic near Sugar Grove, Virginia that I had to decline. I had heard of a small grocery/filling station that served pizzas. I was fixated on taking a break, so I caught a ride into town with a nice family vacationing in the area. As I relaxed in a shaded area of the station enjoying an excellent pizza, about 40 Harley Davidson motorcycles roared into the parking area. They were followed by a red Jeep with the rag top down. A young lady was standing up in the Jeep filming the procession. I guess it was a Harley club headed to Washington, DC to participate in the Rolling Thunder event being held that Memorial Day weekend. I mingled with the bikers and reminisced about the many nights I had spent in the operating room patching up motorcyclists who had lost battles with cars or telephone poles. I had been an orthopedic surgeon for about 25 years in the Indianapolis area. Usually, the motorcyclist was not at fault. Suddenly this very nice lady thrust out her hand toward me holding a wad of money. Since I had been on the trail about two months, my beard, hair, and general appearance had become pretty nasty. I'd lost about 30 pounds by this time. My clothes were dirty, worn, and ill fitting. She thought I was a homeless bum, so she had taken up a collection for me from the bikers!

It took me a moment to understand her kind gesture. I laughed, returned the money, and gave her a brief hug. I explained my appearance was because I was a thru hiker, not a vagrant. When the bikers resumed their journey, I did accept a ride back to the trail in the jeep, which was now leading the procession. I stood by the road waving at my new friends as they sped past. It struck me how I had gone from a seemingly homeless bum to 'leader of the pack' in a few short minutes.

My favorite trail angel was Kerry Smithwick, also known as Scribbles, a lady from Georgia, now North Carolina, that had completed a thru hike in 2011. We had connected via email as I had many questions about my impending hike. She patiently answered them all. She even offered to meet me at Amicoala Falls State Park, near the start of the trail, and hike with me for four days. We had never met and only spoke briefly once by phone. She and her husband Garry were so kind and generous. He dropped us off at the parking lot near the southern terminus of the Appalachian Trail on Springer Mountain, Georgia.

Scribbles was a constant source of good information about trail life that was not available in the books I had read. She showed me flat places to pitch my tent, good water sources, encouraged me to eat, rest with a 'pack off -- shoes off' break, so we could soak our feet in a creek, and other types of information that personalized the hike for me. Remember, I had never slept in my tent until the hike started, and had never actually cooked and eaten a trail meal. I attribute a huge part of my success to her guidance and encouragement. She got this old man pointed in the right direction. She got me off to a good start and made me think my goal of hiking the whole AT was actually possible. I'll always be indebted to Scribbles. She's a lifelong friend.

I met her again the following spring near Asheville, North Carolina, when I returned to do some trail magic in April 2013, along with Wildflower, another former through hiker I had met on the trail in 2012. Wildflower and I served 34 hikers as they came out of the Smokies and crossed under interstate 40. It felt very good to be giving back. The next day Scribbles met us at a rib joint in Asheville for lunch. It was great to catch up with her. She looked fantastic!

Churches are famous for their generosity along the trail, providing trail magic in many forms. The First Baptist Church in Franklin, Tennessee picked up all hikers who wanted free food from the local hotels and delivered us to their fine church for breakfast. They had a large poster for us to sign as well as posters from previous years. I was able to find signatures from former hikers I'd heard about. They also took our pictures, applied them to a postcard, and let us address the card to anyone who might like to receive a message from us along with the picture. That was a happy time -- Franklin First Baptist sure gave us a nice break from the trail. Many church members were involved with helping the hikers. They seemed to be having a good time too. This congregation took good care of Gipcgirl, a contributor to this book, who broke her ankle on the trail in this area a couple of years earlier.

Emma and George gave me a ride into Waynesboro from the Blue Ridge Parkway and delivered me to the Grace Lutheran Church. It was a Wednesday and the church was holding its regular pitch-in in its basement. All hikers were invited. This church was so generous to us with loads of good food. It was fun to talk with the church members and share their excellent meal. That night there were about 15 of us staying in the basement. The church members provided each of us with a cot, snacks, food, advice, coffee, and ice cream. We couldn't spend the day in the basement, because of church activities, but the town library is nearby and a great diner called Weasie's was only a few blocks away. I took a zero there. A hiker named Santa's Helper was the person in charge of the hikers. He gave me a ride back to the trail in the morning so I could continue my hike at the start of the Shenandoah National Park.

Several of the hikers I met at this church I did not see again until I was SOBO in the 100 Mile Wilderness. These included Gingersnap, Peter Pan, and Xango, to

name a few. I got a ride from the Big Meadows wayside
to the lodge at the Shenandoah National Park with a
young couple from Sweden. We had only gone a few
hundred yards when the car stopped suddenly and the
lady vomited after hurriedly opening her door. She had
learned two days earlier that she was pregnant for the first
time. They were excited to return home to tell family and
friends the good news. It seems I was the first one to
know their secret and witness the consequences.

 In Cheshire, Massachusetts I stayed at the St. Mary of the
Assumption Church. There are two rooms off the dining
room/recreation hall that are made available to hikers.
There's no shower or laundry, but a bug free place on the
floor, a roof over your head, a toilet, running water and
electricity were much appreciated. About eight of us hung
out at the church. Atlas, a hiker who carried a guitar the
whole way, played and sang. He actually was quite good.

Another church hostel was the Saint Thomas Episcopal
Church in Vernon, New Jersey. I met up with Gumby and
the Troverts, X and N, again. The Troverts are also
contributors to this book. Some section hikers were also
there named ENT, Preacher Man, and Coach. (Coach is
also from Noblesville, Indiana, my home town. It was
surprising to meet someone who lived where I had grown
up. He'd been a teacher in Tipton -- a smaller town near
Noblesville -- where I currently live. We even had a
mutual acquaintance, John Martin, another teacher. About
a year later, Coach, John, and I hiked together for several
days in southern Indiana.) The Episcopal church let us
sleep on the floor and use the shower and kitchen. It was
near the center of this small town so fast food and
resupply were convenient. They had a chore box filled
with 3 x 5 cards. Hikers would randomly draw one out
and were expected to do that 'chore' as payment for
staying at the church. My job was to vacuum the whole
downstairs. I was glad to do it.

Words of Wisdom

Buy copies of *AWOL's AT Guide* and give them to those friends/relatives who will be following you on your hike. If they will be mailing you something, you can give them the date you'll likely be at a particular location as well as the page number that gives info about the address. Never mail your stuff to a post office, because they have restricted hours and might be closed. A better choice is a business or hostel that will accept items for you and be open most of the time. This will help coordinate the reunion of you and your stuff. Most of your communication from the trail will be in the form of a text. In town or near interstate highways your phones are more likely to work. I used Verizon and was pleased with it. If you are thinking of losing weight try the AT Diet. You can eat as much of anything you want, but you have to carry it around with you for several days first. It can't require refrigeration. You also must carry the stove, pots and pans, as well as the utensils you'll need. You'll have to be able to force all of that stuff into a small water proof bag.

Keep the stuff in your pack dry! This is critically important. If it's wet, it's heavier and won't keep you as warm. Even in warm weather sleeping in a wet bag is no fun. Read the registers found in shelters. It's your only way to communicate with those behind you and learn about those ahead. South bounders are often a good source of general information about the trail ahead of you. The weight you carry is actually very important. Excess weight does wear you down. However, having all you need can be life-saving. It's an unending game you'll play with the trail. Do I have all that I really need but nothing else? Food and water are critically important but do you need another pair of underwear, several pairs of socks, more than two shirts, or more than two pairs of shorts?

Many hikers carry a comfort item, a pillow, Kindle, journal to write in (guilty), or pictures of their grandkids (guilty again). I wondered why, after I had lost 40 pounds or so, the hike wasn't much easier?

A good idea is to have some business cards made with your contact info on them. You can give these to the people you might want to keep in touch with whether they be hiker, trail angel, hostel owner, etc. I can recall hundreds of trail names but almost no 'real' names. Part of the mystique of the trail is to shed your pre-hike identity and start fresh with a new name, a new identity. The trail is a good place to hide. You can lie to me all you want and I'll believe you if you share your Snickers.

Rest when you can. Taking a day off now and then is healing, both mentally and physically. It's not a race. The first one to Katahdin loses. Praise every trail maintainer you meet. These people work hard to make your hike a safe success. They are volunteers. They love the trail for all the right reasons and represent what is good and decent about this great country.

Stay well hydrated. The electrolyte drinks really do have some benefit. I thought water-based drinks would be all I needed, but, about the time I got to Vermont (I can be a slow learner), I discovered Gatorade really was helpful. One of my medical school professors helped develop it as a fluid replacement drink for the Florida Gators football team.

Remember always, it's the journey, not the destination. There will be times you will want to do many miles a day because you're meeting someone, you're behind schedule, bad weather, etc. That's okay, but there were times I felt like a human speed bump as the youngsters raced past.

Their trail experience was different from mine. Speed obscurcs beauty. Nevertheless, hike your own hike!

Start slowly. Cranking out the miles in the beginning is hard because Georgia is tough! Remember, the best way to get in shape to hike the trail is to hike the trail. You'll get your 'trail legs' in 4-5 weeks. I did many Zumba classes and step aerobic classes every week before I left. Sometimes I wore my hiking boots to help break them in. Being surrounded by attractive sweaty women in tight clothes was also motivation! My Zumba family was supportive as the old man headed toward Maine. Before I left for my hike there was a "Goodbye Doc" night at class. Thank you Tricia Harlow for being a great Zumba instructor and quality person!

The Beginning of a Day

My typical hiking day started well before daylight. Long before dawn there is that lonely solitary bird call off in the distance, faint and questioning. Time passes, then it calls again, followed by silence. Eventually another bird answers, perhaps closer, but this one sounds inpatient. Soon, several birds add their opinions about the promises of the day, the likelihood of finding food, or perhaps some interesting distraction. When the light is enough to silhouette the trees the birdcalls become a chorus as the excitement builds. I'm surrounded by their music. The noisier they become the greater the likelihood of a fair weather day. I'll warm some water for a cup of instant coffee and enjoy it as I write in my journal, check my cell phone to see if I have any reception, study *AWOL's Guide*, then warm some more water.

Breakfast includes instant grits, cream of wheat, or oatmeal. Once you get to Pennsylvania grits are rarely available. It's a southern thing I guess. Then maybe I'll have another cup of coffee before I emerge to pack

everything up and move generally north east. I liked to
sign the shelter registers "NE on the AT is NEAT!"

Like the birds, the start of my day is typically filled with
anticipation, even suspense at what the trail may bring.
No two days were ever alike. There was always the
feeling of fascination and mystery that gave an edge to the
day. What was today's hike going to be like? Did I have
enough food, enough water? Was the weather going to be
helpful or make the hike more difficult? Did I have
everything I needed? Were my family and friends okay? I
hated it when there was an extended rainy streak. It
seemed it rained for two weeks straight in Virginia.

A Day Without Sunshine

A day without sunshine
Is like a kiss without lips
A geek without WiFi
A girl without hips.

The trail likes the rain
The slippery slopes
The rocks, the roots
It's a magnet for dopes.

Toe grabbers will get you
Big ups, big downs
The bugs, the snakes
All welcome us clowns.

The vistas are socked in
All rain, no sun
Then how is it possible
I'm having so much fun?

Soap? Who needs it?
You'll get dirty again.

The smells, why bother
They just show where you've been.

Then the sun comes out
All is forgiven.
The flowers, the friends
The trail is for livin'.

What does the trail teach?
What is the draw?
That if you stay focused
You can enjoy it all.

So when the day comes
That your hiking is thru
Reflect on your victory
There's nothing you can't do!

Go change the world
One step at a time.
Left, right, left, right.
Time to end this rhyme

Not a day went by that I didn't say the Lord's Prayer and
a brief personal prayer for the other hikers, the trail
angels, and even for myself. I always felt guided and
protected. While I was alone -- possibly 80% of the time I
spent hiking -- I never felt lonely. God's presence felt so
real that I occasionally stopped hiking and looked behind
me to see if someone was actually standing in the woods
watching me. I learned to enjoy that feeling of guidance,
approval, and protection after I had been on the trail two
or three months. I had endless verbal and mental
conversations as I hiked along. Any person hearing me
would think I was talking to myself, but it did not feel that
way to me. This feeling of spirituality left when I went
into towns or stayed in a shelter with others.

An average day in the early part of my hike was an interesting study in the meaning of time. Yes, the hours passed as the Sun moved across the sky, but the seasons seemed to pass as I ascended and then descended the mountains. In the valleys, or gaps there was the new growth of Spring, but on the summits it was still Winter-like. As I walked further north these changes became more obvious as the days passed. Sometimes it felt like I walked from Summer to Spring and then Winter in a single day, as I went from valley to summit.

I typically hiked until late morning, then stopped for a 'pack off' rest, ate something, maybe took a nap, then loaded up and resumed my hike. Another break might come mid afternoon. My final stop occurred when I was tired and hungry and saw a flat spot in the woods. I always tried to pitch my tent so it was facing east. I enjoy watching the new day find me. Beginnings not endings are my favorite, whether it be a day or a relationship. I would much rather be an obstetrician then a mortician. I delivered a few babies early in my medical career. The cry of a newborn is sweet music. After delivering a baby and seeing what the mother went through, I always wondered why anyone became pregnant again. It felt the same when I heard about someone who had hiked the AT more than once. Must be hiker amnesia.

So, as a day of hiking, contemplation, and experience of natural beauty comes to a close, I crawl into my tent, into my bag, check to see if I have a signal, make sure things I might need are where I can find them in the dark, then fall asleep listening to the night sounds and the promises they make to me of yet another sunrise. After being on the trail for a few weeks I could no longer recall the boredom of being alive.

Since the privy was an integral part of the trail experience, another poem came to me, or from me?

290

Privy Woes

When memory keeps me company
And moves me to smiles or tears
A weathered trail through fourteen states
Beckons through the years.

Trail angels give their food to you
Always generous with the feast
The trail will give you what you need
A privy not the least.

I did my 'duty' promptly
Never having to use a cob
But the torture of an ice cold seat
Might make a Marine sob.

I remember the Jenkins shelter
It's privy had no walls
Splinters, wasps and spiders
Were always a threat to my b---s

And Tennessee, you had none
What's the deal with that
I had to find an OK place
And bury it like a cat.

Eventually I would sympathize
I found the truth I sought
As to why my dog would circle
Till she found just the right spot.

Silence on the Trail

Silence on the trail has many meanings. There are many
kinds. There's the hush before a storm. The birds become
quiet, the wind stops, the forest seems to hold its breath.
Then the distant rumbles of thunder begin and come
closer as the wind picks up. This forest noise has an air of
anticipation, an empty silence that precedes the coming
roar. With the passing of the storm there's a sound of
relief from the woods and joy it must feel in surviving
once again this cycle of life. Then there is the silence that
welcomes sunrise. The birds have announced the coming
day but they become quiet as the sunlight begins to reach
the mountain tops, then the nearby trees, and finally the
rocks and ground in front of me. There's also the quiet,
the hush that says goodbye to the day as the Sun escapes,
dragging the light with it. This silence has a certain
finality. It seems to be saying goodbye, but at the same
time welcomes the night as we might welcome a good
friend. There's also the silence you 'feel' when you are
alone in the woods. It is a therapeutic stillness, a form of
peace that moves you out of your self awareness and into

the primal bond you have with the woods, with the moment. I miss the many hours I would spend alone every day. This is a simple quiet that heals and justifies you in a spiritual way. Somehow it seems cleansing. The trail is less about creating your identity than it is about finding it. Nobody can hike for months without learning more about themselves than they sought. This knowledge comes with the speed of rust. Hiking requires so very little mental capacity. You are so firmly in the present moment that the past and future seem closer. One day the same old questions have different answers.

In closing I think the main draw of the trail is the freedom that comes with it, to do, to be, to leave behind, to look ahead. Just as 'vision impaired' people develop heightened awareness in other senses, time on the trail causes the hiker to become more keen on the things they're surrounded by whether it be the natural beauty, the individualism of other hikers or their good fortune to be on the trail hiking their "own hike".

~ ~ ~

I'm a retired orthopedic surgeon living in the central Indiana area. I was able to thru hike the trail in 2012, mostly because of good luck. I had no hiking experience but was in pretty good shape because of the many zumba and step aerobic classes I took weekly. At 66 I was the oldest thru hiker I met that year.

Eight of my ten grandkids had birthdays while I was on the trail. In spite of the sporadic cell phone service I was able to call each one and sing to them. This may not mean much to most of you, but to me it's an example of the ever present serendipity on the trail.

Chapter 15 The Appalachian Trail: My Story

The Trail Speaks

Ever wonder what I think? What I feel? I'm the Appalachian Trail and I'll tell you my history, my opinions and my feelings. This is my story. Clearly I have witnessed many things, as many as there have been people walking, working or picnicking on me. I must say I enjoy day hikers the most. They are just happier with children, pets, picnics, good smells and clean clothes. My next favorite group is the trail maintainers. They work hard, volunteer their time and sweat to take good care of me. Over the years I've grown about 200 miles. They've changed my actual route many times, built steps, removed trees, and taken me away from roads and towns when they could. I appreciate that. I was never meant to be paved. My next favorite group are the section hikers. They seem to enjoy their limited time with me, are not always worn out, stop to appreciate my beauty, and look forward to returning again to hike and experience another part of me. Then there's the thru hikers. I know them all. I admire their courage but also recognize what it takes to just keep moving. In some ways thru hiking the trail is like going to church. You may not always act right or deserve grace but receive forgiveness often enough to justify the effort, the result.

Neels Gap ... the only place where I am under a roof

Hikers have different styles and gear. I've noticed that more of them use hammocks, which I consider a lazy person's throne. Nobody ever cries in a hammock. My relationship with long distance hikers is like a marriage. Forgive the difficult times and enjoy the unexpected rewards.

Now, for some history. I began in the mid 1930's to appreciate the human efforts to create me. My father must be Benton McKay, who in 1921 thought of and published an article describing a trail connecting farms and work-study camps along the eastern continental divide. My first section opened in New York in 1923. Myron Avery took over the leadership in 1932 and led efforts that completed me on Sugarloaf Mt. in Maine in 1937. Ongoing rerouting continues today. Originally I did not include Roan Mountain in North Carolina, Grayson Highlands in Virginia, Nuclear Lake in New York, or Saddleback Mountain in Maine. I was only 2,044 miles long in 1948.

The PCT was not completed till 1993 making it 56 years younger than me.

The Civilian Conservation Corps was brought in to help build and repair me in the 1930's, especially in the Smoky Mountains and in the Shenandoah National Park. I originally climbed straight up and down mountains leading to erosion and a rough hiking surface. I was a lot smoother when I was young. Aren't we all? I am maintained by the volunteers of 31 hiking clubs along my length. I am managed by the National Park Service, United States Forestry Service, and especially by the Appalachian Trail Conservancy. I spend the vast majority of my time in wild lands. I'm located only briefly in towns, on roads, or farms.

I am the longest "hiking only" trail in the world. Each year I have about three million visitors. The Appalachian Trail Conservancy does an excellent job of protecting, maintaining, and promoting me. 1968 President Johnson signed the National Trails System Act protecting me and my sister, the Pacific Crest Trail, from development by designating us as the first National Scenic Trails. The Continental Divide Trail is still a work in progress. All but a few miles of me are owned by the federal government. The owners of the private land are very generous to allow its use. This makes me essentially a 2,200 mile long 1,000 foot wide national park maintained by volunteers. Bicycles are only permitted along the C&O Canal in Maryland and on the Creeper Trail in Damascus, Virginia. Horses can only be used on a small section in the Smokies.

Changes in altitude and latitude cause hardwoods to be the primary trees in the south and coniferous trees in the north. The first part of the trail above tree line is Mt. Moosilauke in New Hampshire. The Franconia and Presidential range, including Mt. Washington, run for 13 miles above tree line and is the largest continuous Alpine environment in the United States east of the Rockies.

I originally started at Mount Oglethorpe in northern Georgia but because of nearby overdevelopment I was moved 20 miles away to Springer Mountain in 1958. The last remnant of the original trail from Mount Oglethorpe is the 8.5 mile approach trail that runs from Amicaola Falls State Park to my current start on Springer Mountain.

My most photographed tree

Now back to my real treasure, the hikers. While it is true you should not judge others when you have no idea what their life has been like, I act as a great equalizer, a form of forced camaraderie. All thru hikers eventually learn what life on the trail means. While so many hikers obsess over pack weight, trying to get rid of whatever isn't useful, I think they might improve their lives by also eliminating whatever isn't beautiful or joyful. Rejoice in the moment. While thru hikers are not my favorites they are the most needy. I try to protect them, help them make good choices, and to learn from their mistakes. Most of them

are in transition --- recovering from a divorce, an
untimely death, a lost job, or recently out of college or the
military, and are looking for a new direction. This trek is
less about figuring things out than just discovering what
was there all along. Some say I am like a large intestinal
track, feeding on innocent thru hikers at the start and
digesting them until they become hardened 'veterans' in
the end. I have seen many relationships develop into
romance then deteriorate. I am like your favorite teacher
in school. I will work you hard but you will learn more.
When people have hiked me extensively, leave for a time,
then return, I proudly hear them say, "It feels like coming
home." I enjoy waking up from winter's quiet to the
rebirth of the woods around me and to meet the new
hikers, workers, and pilgrims who enjoy me as I am.

wisdom from THE trail

Don't judge me by my relatives
The PCT welcomes mules!
The Continental Divide Trail
Meanders about like a fool.

To all my hiker friends out there
This should be understood
Your life won't always be fair
But it can always be good.

Live so that when you die
Only a part of you is gone.
No matter how good or bad things are
There'll always be another dawn.

Change is inevitable
So welcome it as certain.
Expect the coming miracle
Behind the next curtain.

In the end my hikers learn
The best is yet to come
No matter their direction
Or wherever they are from.

I challenge you, I dare you to quit.
The solitude will force you to think.
So when in doubt just take the next step
From hell to fun I make the link.

Your life is rushing to the day
When silent as a mime
You find it was but a single beat
Within the heart of time.

Glossary

Aqua Blazing   Refers to the act of using a boat to travel the AT. This is most often tried on the Shenandoah River in the Shenandoah National Park.

AYCE   An acronym for an All-You-Can-Eat buffet. These are a hiker's favorite commercial places to eat.

Bald   The treeless tops of some mountains in the south not associated with being above tree line.

Base Weight   The total weight of your gear not counting water and food.

Bear Bag   The waterproof container that you keep your food in and suspend from tree limbs to keep out of the reach of bears.

BiBo   An abbreviation for going both ways on the trail. Similar to a flip-flop but more descriptive.

Blaze   A trail marker. Comes in many colors.

Blue Blaze   Signifies a trail that's not the AT.

Boomerang Zero   When you walk the wrong way, realize your mistake and return to where you started. You may have walked many miles but progressed none. Very frustrating!

Bush Whacking   Walking off the marked trail. Can be damaging to the forest floor.

Cairns   A pile of rocks, usually pyramidal shaped, that marks the trail. Often there's a blaze painted on it.

Camel Up   Refers to drinking a little extra water so you don't have to carry as much in a container.

Croo   Summer staff that works in the AMC huts in the White Mountain range.

Flip-flop   Hiking the trail in one direction then moving ahead to hike back to the point where you left.

Gap   A southern term referring to what might also be called a notch, pass, valley, etc.

Gorp   Another acronym for trail mix meaning Good Old Raisins and Peanuts.

Hiker Box   A box found at hostels, motels or other places where hikers gather that is filled with discarded gear, food, books, etc. that might be coveted by somebody.

Hiker Midnight   Usually around nine.

Hut   AMC controlled hostel in the White Mountain range.

HYOH   Another acronym that stands for Hike Your Own Hike. This is a guiding principle on the trail.

Lean To   Shelters in New England.

Leave No Trace   The commandment for all who use the trail that they keep it clean.

Nero   Hiking an abbreviated mileage day.

NoBo   Someone who is hiking northbound.

Pink Blazing   Someone who is possibly off the trail following a romantic interest.

Privy   Also called in some areas an outhouse. Frequently found near shelters. Tennessee doesn't have any.

PUD   Pointless Ups and Downs. Usually discussed after one has hiked the 'roller coaster' in northern Virginia.

Puncheon   Rudimentary wooden bridges over bogs and marshy areas. Can be very slippery. Found most often in NJ, VT and ME.

Purist   A hiker who insists on walking past every white blaze with a full pack.

Resupply   Purchasing supplies needed in town. Occasionally the calories per penny is used as a measure of the value of a food item.

Section Hiker   A hiker who may do parts of the trail at a time possibly with the goal of doing the whole trail eventually.

Shelter   Usually a three sided building with a roof and a floor for the use of hikers who choose to not tent. Often they are arranged about a day's hike apart.

Slack Packing   Carrying a small pack such as a day pack that has only the essentials in it. You must eventually return to reclaim your complete pack but hiking with the decreased weight is invigorating for some.

SoBo   A hiker who is primarily headed south. When referring to a thru hiker this means they start in Maine and head for Georgia. Usually they start early June and hope to be done between Thanksgiving and Christmas.

Springer Fever   That 'illness' that calls hikers back to the excitement of their attempt at a thru hike, beginning at Springer Mt.

Stealth Camping   Finding a place to camp along the trail away from shelters and campsites that is considered nontraditional. Occasionally these are in forbidden areas.

Thru Hiker   Someone who has hiked the entire AT in a twelve month period.

Trail Angels   They come in all shapes and ages. People who aid hikers with food, rides, or any other form of kindness. Critically important to enhancing the 'trail' experience. An American phenomenon.

Trail Magic   Those gifts of food, aid, encouragement provided by trail angels.

Triple Crown   The three longest and best known hiking trails in the USA. They are the Appalachian, Pacific Crest and Continental Divide trails.

Vitamin I   A trail name for ibuprofen, a commonly used and needed analgesic and anti-inflammatory.

White Blazing   Hiking the Appalachian Trail.

Yellow Blazing   Traveling by either hitchhiking or road walking.

Yogi-ing   Term referring to begging for food but not really seeming to. Works best at picnic areas filled with families.

Yoyo   Refers to when a hiker has thru hiked the AT then turns around and hikes back. Considered by some to be a rare form of 'bipolar' behavior.

Zero   When a hiker takes the day off to recuperate, resupply, eat town food, do laundry, bathe.

Acknowledgement

I want to sincerely thank the several contributors to this book, along with Jeff Rasley, my Himalayan hiking-friend, who helped me edit this book. Each one has been a pleasure to work with. I mean that! I've hiked with most of them, met several more off the trail, and I'm looking forward to meeting those I have not met in person. The trail is our common denominator, uniting us with shared experiences that may be separated by time but not emotion.

Jim Dashiell MD aka Funnybone

Made in the USA
Lexington, KY
13 March 2017